PreK-5

W9-CNC-460

The

FRUGAL
Science
Teacher

STRATEGIES AND ACTIVITIES

PreK-5

The FRUGAL Science Teacher

STRATEGIES AND ACTIVITIES

Linda Froschauer, Editor

NSTApress

National Science Teachers Association

National Science Teachers Association

Claire Reinburg, Director
Jennifer Horak, Managing Editor
Andrew Cocke, Senior Editor
Judy Cusick, Senior Editor
Wendy Rubin, Associate Editor
Amy America, Book Acquisitions Coordinator

SCIENCE AND CHILDREN
Linda Froschauer, Editor
Valynda Mayes, Managing Editor
Stephanie Andersen, Associate Editor

SCIENCE SCOPE
Inez Liftig, Editor
Kenneth L. Roberts, Managing Editor
Janna Palliser, Consulting Editor

ART AND DESIGN
Will Thomas Jr., Director
Tim French, Cover and Interior Design

PRINTING AND PRODUCTION
Catherine Lorrain, Director

NATIONAL SCIENCE TEACHERS ASSOCIATION
Francis Q. Eberle, PhD, Executive Director
David Beacom, Publisher

Copyright © 2010 by the National Science Teachers Association.
All rights reserved. Printed in the United States of America.
13 12 11 10   4 3 2 1

Library of Congress Cataloging-in-Publication Data
The frugal science teacher, prek-5 : strategies and activities / edited by Linda Froschauer.
    p. cm.
 Includes bibliographical references and index.
 ISBN 978-1-936137-01-5
 1. Science--Study and teaching (Elementary)--United States. 2. Science--Study and teaching--Activity programs--United States.
I. Froschauer, Linda.
 LB1585.3.F78 2009
 372.3'5044--dc22
                              2009044630

 eISBN 978-1-936137-80-0

CONTENTS

PART 3.   TEACHING STRATEGIES THAT MAXIMIZE THE SCIENCE BUDGET

PART 4.   INSTRUCTIONAL LESSONS THAT MAXIMIZE THE SCIENCE BUDGET

Preface

by Linda Froschauer

frugal *(froo'gal) adj. Practicing or marked by economy, as in the expenditure of money or the use of material resources. See synonym at* sparing. *2. Costing little; inexpensive.*

—The American Heritage Dictionary

Frugality practically defines how we as teachers approach provisioning our classrooms. (I half expected to see a picture of a science teacher next to the entry!) We cleverly create learning opportunities with limited resources and have amazing aptitudes for stretching shrinking funds and doing more with limited resources. Still, we find ourselves augmenting school and district funds with our own dollars, digging into our own pockets to purchase equipment and other essentials. A quick web search suggests that K–12 teachers spend between $475 and $1,500—per year—on classroom materials. And we do this willingly because we know it makes a difference in our students' learning.

In an issue of *Science Scope* devoted to limited classroom resources, editor Inez Liftig expressed concern about giving tacit approval to the expectations that teachers should spend their own money to outfit their classrooms: "I wanted to be very sure that we did not send the wrong message about whether or not science teachers should spend their own money to support instruction. . . . Parents and school districts should not expect teachers to pay for equipment and supplies from personal funds, and we should not have to choose between doing them at all" (Liftig 2007, p. 6). I share her concern, but my intent here is not to lecture or opine. Rather, I hope this volume provides a valuable reference at a time when we all need to be resourceful.

To collect all of the articles, books, websites, and organizations that can help you save money is an impossible feat. Not only is there a tremendous quantity of available resources, but the information also changes rapidly and is best pursued through internet searches. Therefore, you will not see lists of websites, grants, and "free" opportunities in this book. Rather, you will find a collection of inspiring articles and book chapters that will provide you and your students with valuable, standards-based learning opportunities that can also serve as springboards to additional investigations. The authors detail untapped resources for materials, reimagined uses for items you already have at home or school, inexpensive workarounds to costly classroom projects, and creative activities that require only free or inexpensive materials.

In addition, many articles and chapters include suggestions for further reading that may expand on the ideas discussed, apply a similar learning tool in a different way, or revise a particular activity for use with different grade ranges. These additional resources are available through the NSTA Science Store (*www. nsta.org/store*), for free or little cost.

A WORD ABOUT ORGANIZATION

This book comprises five categories, or overarching strategies, for thinking about how to conduct science investigations without spending a great deal of money—either your own funds or those acquired through your district budget.

Student-Created Constructions

When students build their own equipment or create their own models, they have a greater connection to the overall experience, thus enhancing learning. An amazing number of investigations can be developed with a single piece of paper, throwaway items, or dollar-store finds. You already may be familiar with more traditional student constructions, such as paper airplanes, and the lessons they convey. Think how much more students could learn from building roller coasters or paper towers.

Teacher-Created Constructions and Repurposed Materials

Science teachers are great savers of materials. We check out sale bins in stores and rinse out used containers. We collect soda bottles, aluminum cans, shoe boxes, scraps of wood, odd lots of rubber bands, old CDs . . . anything that may possibly be useful in our classrooms. This section suggests ways to put those materials to good use in two general categories: repurposing materials that we have collected and building equipment for student use from free or inexpensive components.

Teaching Strategies That Maximize the Science Budget

There are many ways to reorganize our instructional approaches that enable quality learning to occur at reduced cost. The articles in this section provide suggestions on how to engage students through a variety of strategies. Although the strategies are explained within the context of a specific content area, they can serve as creative inspiration as you consider how to adjust lessons in *any* content area. Creating project materials, playing games, drawing cartoons, developing class newsletters, using learning stations, and tapping into current events all require minimal financial investments but provide enriched experiences for students. Many of these ideas also integrate other subject areas to provide broader curriculum impact.

Instructional Lessons That Maximize the Science Budget

The fourth section offers a collection of life science, Earth science, and physical science and chemistry investigations. They are specific to a given content area but utilize materials that may stimulate ideas for innovative activities with any subject matter. You can use them as they are or modify them to fit your curriculum. Several articles highlight the use of outdoor spaces around your school site that are ideal for scientific investigations.

Funds and Materials

Even after implementing the ideas in this book, you may still have classroom needs that prove too costly to be fulfilled through your budget (or pocket). This section presents suggestions for how to acquire those additional funds.

ADDITIONAL RESOURCES FROM NSTA

In this volume, I have culled some of the most useful NSTA print resources for maximizing your classroom dollars. However, NSTA also provides a variety of free electronic resources that are available for members and nonmembers alike to improve both teaching and learning.

e-Publications

Individual articles from *Science and Children, Science Scope,* and *The Science Teacher,* as well as chapters from NSTA Press books, are available in electronic format from NSTA's online Science Store (*www.nsta.org/store*). Many of these—at least two articles per journal issue and one chapter per book—are free to everyone. (The balance of articles is free to NSTA members and available for a small fee to nonmembers.)

Teachers and administrators can also keep up with what's happening in the world of science education by signing up for free weekly and monthly e-mail newsletters (*http://www.nsta.org/publications/enewsletters. aspx*). *NSTA Express* delivers the latest news and information about science education, including legislative updates, weekly. Every month *Science Class* offers teachers theme-based content in the grade band of their choosing—elementary, middle level, or high school. News articles, journal articles from the NSTA archives, and appropriate book content support each theme. *Scientific Principals,* also monthly, provides a science toolbox full of new ideas and practical applications for elementary school principals.

Learning Center

Anyone—teachers, student teachers, principals, or parents—can open a free account at the NSTA Learning Center, a repository of electronic materials to help enhance both content and pedagogical knowledge. By creating a personal library, users can easily access, sort, and even share a variety of resources:

Science Objects are two-hour online, interactive, inquiry-based content modules that help teachers better understand the science content they teach. New objects are continually added, but the wide-ranging list of topics includes forces and motion, the universe, the solar system, energy, coral ecosystems, plate tectonics, the rock cycle, the ocean's effect on weather and climate, and science safety.

SciGuides are online resources that help teachers integrate the web into their classroom instruction. Each guide consists of approximately 100 standards-aligned, web-accessible resources, accompanying lesson plans, teacher vignettes that describe the lessons, and more. Although most SciGuides must be purchased, there is always one available at no charge.

SciPacks combine the content of three to five Science Objects with access to a content expert, a pedagogical component to help teachers understand common student misconceptions, and the chance to pass a final assessment and receive a certificate. Yearlong SciPack subscriptions must also be purchased, but one SciPack is always available for free.

Anyone may participate in a live, 90-minute web seminar for no-cost professional development experiences. Participants interact with renowned experts, NSTA Press authors, scientists, engineers, and other education specialists. Seminar archives are also available on the NSTA website and can be accessed at any time.

Particularly popular web seminars are also offered in smaller pieces as podcasts that can be downloaded and listened to on the go. These 2- to 60-minute portable segments include mini-tutorials on specific content and ideas for classroom activities.

Grants and Awards

NSTA cosponsors the prestigious Toyota Tapestry Grants for Teachers (*www.nsta.org/pd/tapestry*), offering funds to K–12 science teachers for innovative projects that enhance science education in the school or school district. Fifty large grants and at least 20 mini-grants—totaling $550,000—are awarded each year. NSTA also supports nearly 20 other teacher award programs, many of which recognize and fund outstanding classroom programs (*www.nsta.org/about/awards.aspx*).

You may not save hundreds of dollars a year by following the recommendations found in this book. You will, however, find creative ways to keep expenses down and stretch your funds while building student understanding. . . . And perhaps you will be inspired to invent your own low-cost constructions, develop even more inexpensive student activities, find additional uses for everyday items, or uncover a wealth of new resources for obtaining classroom materials.

Reference

Liftig, I. 2007. Never cut corners on safety. *Science Scope* 30 (6): 6–7.

PART

1

Student-Created
Constructions

Chapter 1

Roller Coasters

edited by William C. Ritz

INVESTIGATION

Observing how gravity affects marbles as they roll

PROCESS SKILLS

Observing, comparing

MATERIALS

Masking tape, pipe insulators (¾ in. or larger), various size marbles

PROCEDURE

Getting Started

Cut pipe insulators in half lengthwise. Demonstrate how to connect the pipe insulators with masking tape.

Further Reading

- "Roller Coaster Inquiry," from the September 2004 issue of *The Science Teacher*

Next, show how to connect the tracks to create loops, ramps, and curves. Encourage children to drop marbles on the tracks to discover how gravity affects the marbles' roll. Give children time to explore on their own before asking questions.

Questions to Guide Children
- What do you think will happen if we change this loop (ramp, curve)?
- Can you make the marble go faster? Farther?
- What will happen if we add another curve at the bottom? At the top?

What Children and Adults Will Do
Children will first use the structures that adults have set up. After they have enjoyed using those structures for a while, encourage children to change some of the component pieces or to start from the beginning and create their own. The marbles will roll an amazingly long distance so you may want to put up some barriers around the perimeter.

Closure
Tell children the materials will be in the block area so they can continue to try out their ideas.

CENTER CONNECTION

Put materials in either the block area or another area where there is space to build. Take the materials outside if you have a carpet or blanket on which to place them so marbles don't get lost.

LITERATURE CONNECTION

Roller Coaster
by Maria Frazee, Voyager Books, 2006.

ASSESSMENT OUTCOMES AND POSSIBLE INDICATORS*

- **Science: Scientific Skills and Methods**
  Begins to describe and discuss predictions, explanations, and generalizations based on past experiences.

- **Mathematics: Geometry and Spatial Sense**
  Builds an increasing understanding of directionality, order and positions of objects, and words such as *up, down, over, under, top, bottom, inside, outside, in front,* and *behind.*

- **Social and Emotional Development: Cooperation**
  Increases abilities to sustain interactions with peers by helping, sharing, and discussing.

- **Approaches to Learning: Reasoning and Problem Solving**
  Grows in recognizing and solving problems through active explorations, including trial and error, and interactions and discussions with peers and adults.

*Source: *Head Start Bulletin.* 2003. Issue No. 76. U.S. Department of Health and Human Services, Head Start Bureau. *www.headstartinfo.org/publications/hsbulletin76_09.htm.*

What to Look for	
Not Yet	Child shows no interest in helping to build a roller coaster.
Emerging	Child watches others build a roller coaster; participates by rolling a marble and retrieving marbles but does not help in building the roller coaster.
Almost Mastered	Child eagerly participates in helping to build a roller coaster; assists in taping several lengths together; tries out different configurations; discusses findings.
Fully Mastered	Child builds at least part of the roller coaster without help; works with others to complete a complex structure; experiments by changing the height and angle of various selections; asks questions about the project; asks for the activity to be repeated.

chapter 1

ROLLER COASTERS
Family Science Connection

Collect several cardboard paper towel and toilet paper centers. Together with family members create a roller coaster track using the cardboard centers. Experiment by making the track curve, loop, and drop from a great height. Next, roll marbles or round objects through the track. Play with your newly created roller coaster by changing the track, adding to it, and timing how fast the marbles move within the track.

Comments or questions that may add a *sense of wonder* to this activity:

- What do you think will roll the fastest on the roller coaster: a marble, a toy car, or a pencil?
- What ended up rolling the fastest and what do you think made it win?

MONTAÑAS RUSAS
Ciencia en Familia

Guarden varios centros de cartón de las toallas de papel y papel sanitario. Junto con miembros de la familia hagan un riel de la montaña rusa usando los centros de cartón. Experimenten haciendo el riel curvo dando varias vueltas, y caídas de altura. Luego dejen rodar canicas u objetos redondos dentro del riel. Jueguen con su montaña rusa cambiando de rieles, agregando más rieles, tomando el tiempo que toman las canicas para pasar por el riel.

Comentarios o preguntas que pueden *despertar curiosidad* en esta actividad:

- ¿Qué piensas que rodará más rápido en la montaña rusa, una canica, un carro de juguete, o un lápiz?
- ¿Qué fue lo que rodó más rápido y qué piensas que le hizo ganar?

This chapter first appeared in A Head Start on Science: Encouraging a Sense of Wonder *(2007), edited by William C. Ritz.*

Chapter 2

Taking Flight With an Inquiry Approach

by Kathryn Silvis

This paper airplane lesson has been used with sixth-grade students to introduce scientific terms and concepts that students need to know before they design and conduct their own inquiry experiments. Terms such as *manipulated/independent variables*, *responding/dependent variables*, and *constants* are defined within this lesson, and students come to understand the importance of changing only one variable in scientific experiments. In addition to science concepts, mathematics skills are embedded in this lesson as students measure paper airplane flights to the nearest centimeter and calculate range, mode, and median in a hands-on way. The 5E model provides a way to organize inquiry-based instruction, and focuses on actively involving students in the learning process (Carin, Bass, and Contant 2005). For this lesson, students engage in creating their own paper airplane model, explore how far their paper airplane will fly, explain the class results of paper airplane flights, elaborate by designing another paper airplane experiment with only one variable, and evaluate patterns found in paper airplane test flight data.

ENGAGE

As students enter the room, they find at their lab tables directions for the lab and supplies for constructing a paper airplane. Each table of four to six students has a variety of different types of paper (such as construction paper, graph paper, and notebook paper), as well as markers, scissors, and tape. After students create their own paper airplane models, they can decorate their airplane as time allows using markers at their tables. There are no specific directions for making the plane, which should ensure a wide variety of models to test during the Explore phase (see Figure 1, p. 8).

Students are then asked to make a prediction of how far in centimeters their paper airplanes will fly and record their prediction in their science journal. Students' predictions can be used as a diagnostic assessment of each student's ability to estimate with centimeters. For example, a student who predicts that his or her paper airplane will fly only 5 cm does not have an understanding of metric estimation. In

Further Reading

- "Ever Fly a Tetrahedron?" from the March 2004 issue of *Science Scope*

addition to each student's personal prediction, students can predict which classmate's paper airplane will fly the farthest distance. This prediction can be recorded on a class data chart.

This activity immediately engages students, who are often excited to make their own paper airplanes in science class. The excitement continues to build as students prepare to explore how far the paper airplanes will actually fly.

EXPLORE

After students have created their paper airplanes and made their predictions, the teacher takes the class to the test flight site. A long, straight hallway works best for test flights, with a metric measuring tape running the length of the hallway. Some hallways have marked sections of tile on the floor, which makes it easier to estimate distances. Because students often shout out with excitement during the test flights, it is best to select a test flight site that will not disturb other classes. The test flights could also be done outside, although wind could add another variable.

After everyone has donned safety goggles, students take turns throwing their paper airplanes and measuring how far they flew to the nearest centimeter, using the farthest points of the planes from the starting line. Students then record their names and flight distances on small sticky notes, which are posted at their landing points on the floor. These sticky notes will be used for a data classification activity during the Explain phase. The sticky notes also serve as placeholders for the planes, so that the paper airplanes can be removed from the landing area to avoid collisions with other paper airplane flights. Students write the distance of their own paper airplane flights in their science journals and compare it to their predictions.

Once all the planes have had their test flights, there should be a set of sticky notes spread out on the floor (see Figure 2), which provides a great opportunity for students to explain patterns in the data.

EXPLAIN

One way to explain patterns in the data is to apply mathematical measures of central tendencies, including range,

Figure 1

Sample paper airplane models

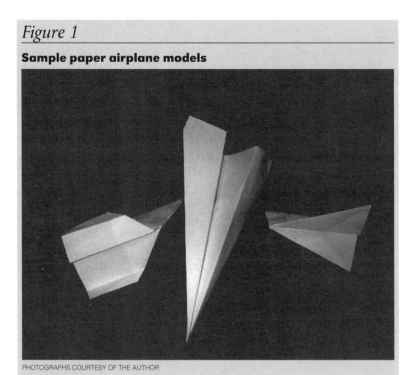

PHOTOGRAPHS COURTESY OF THE AUTHOR

Figure 2

The mode represented by multiple sticky notes

mode, and median. These three terms can be demonstrated in a hands-on way, using the collection of sticky note flight data. The range of the data set will be easy for students to see, as the teacher can direct students to look at the difference between the shortest flight distance and the longest flight distance. After students visualize this difference, they can make the subtraction calculation to determine the actual range (subtracting the shortest flight distance from the longest flight distance). The mode will be clearly evident if any distances occurred more than once because there will be several sticky notes at the same spot on the floor. For example, the flight data shown in Figure 2 shows a mode of 180 cm.

Finally, because the distances are already arranged from least to greatest, the median can be visually represented with a student demonstration. To begin the demonstration, each student stands at the landing point of his or her paper airplane. Then, the student standing at the shortest flight distance and the person standing at the longest flight distance are asked to sit down in unison. This process is repeated until one person is left standing. This person represents the median for the data set, the middle value. If two people remain standing at the end, then the class can determine the middle point between these two distances. After these mathematical calculations are made, students can continue to explain the results of the experiment.

The 5E model helps students develop inquiry skills because students are asked to interpret data to determine *why* something happened, which goes one step beyond simply making an observation (Carin, Bass, and Contant 2005). The questions that a teacher asks at the Explain stage are a critical part of guiding students to think more deeply about experimental results. For this paper airplane experiment, the teacher can ask the following open-ended questions:

- What patterns do you notice in our paper airplane data?
- Which plane had the longest flight? Why do you think so?
- Which plane had the shortest flight? Why do you think so?
- What is the best design for a paper airplane? Why do you think so?

As they respond, students may begin to make some conclusions, such as, "The planes made out of construction paper did not fly as far as the planes made out of graph paper." At this point, the teacher will want to draw students' attention to the many variables that existed in the paper airplane experiment, such

---

Figure 3

Group data sheet for paper airplane flight experiment

What variable has your group selected to test? _____

The variable that your group has selected to test is called the *manipulated,* or *independent, variable.*
What is your *responding,* or *dependent, variable?* _____

All of the other variables in your experiment need to be constant.
What will the *constants* be for your experiment? _____

Explain your group's hypothesis for this experiment—what do you think will happen and why? _____

Describe the steps of your group's experiment:
1.
2.
3.
4.

Results for the experiment:

Independent variable		
Trial 1		
Trial 2		
Trial 3		
Trial 4		
Trial 5		
Range		
Mean		
Median		
Mode		

Explain how your results compared with your hypothesis for this experiment:

_____

_____

as different airplane designs, different people throwing the airplanes, and different paper that was used. The teacher can then introduce specific terms related to variables (Carin, Bass, and Contant 2005) and relate each term to the paper airplane experiment. The *manipulated*, or *independent*, *variable* is the aspect of an experiment that is deliberately changed. This original paper airplane experiment has several independent variables, including airplane design, type of paper, and the force the person uses to throw the airplane. The *responding*, or *dependent*, *variable* is the aspect of an experiment that responds to the deliberate changes made. In the case of this experiment, the dependent variable is the flight of the paper airplane. Finally, *constants* are aspects of an experiment that are deliberately kept the same. The only constant in this experiment would be the paper airplane test site, because everyone tested their airplanes at the same site. All of the other variables were not constant. Because there are so many manipulated or independent variables in this experiment and few constants, it is not possible to make conclusions that focus on just one variable, such as, "The planes made out of construction paper did not fly as far as the planes made out of graph paper." Next, students are asked to fix this problem with the experiment by selecting just one variable to manipulate in the Elaborate section.

ELABORATE

Each small group of four students should design an experiment that manipulates just one variable of paper airplane flights, while keeping all other variables constant. Students can work together to complete the data sheet shown in Figure 3, page 9.

As students are working, the teacher should be completing some formative assessment to ensure the groups are creating experiments that have just one manipulated, or independent, variable. After groups have checked their experimental outlines with the teacher, they can return to the test flight site to test and record the new flight distances. Because students will be completing several trials for each independent variable, they should calculate the range, mode, and median for each set of trials.

EVALUATE

Students can share the results from their small-group paper airplane experiments with the class. Each member of the group should participate in the reporting process. One member can explain the variable that the group decided to change. The second group member can explain the variables that were kept constant in the experiment. A third member can share the results from the test flight trials, including the range, mode, and median for each trial set. Finally, the fourth member can explain whether or not the conclusions from the testing matched the hypothesis. For individual evaluation, students can be asked to explain in their science journals at least two patterns from their small-group experiment data.

CONCLUSION

After students have experienced each section of this 5E lesson, they have a better understanding of how to design a scientific experiment. This lesson also draws students' attention to a common flaw in experimental designs—changing more than one variable. When the paper airplane lesson was used at the start of the year, students had greater success designing their own experiments throughout the year because they understood the importance of changing just one variable while keeping everything else constant. Students truly "take flight" with an inquiry approach as a result of this lesson!

Reference

Carin, A. R., J. E. Bass, and T. L. Contant. 2005. *Teaching science as inquiry*. 10th ed. Upper Saddle River, NJ: Pearson Prentice Hall.

This article first appeared in the September 2008 issue of Science Scope.

Chapter 3

String Racers

by Bruce Yeany

INSTRUCTIONAL INFORMATION

Overview

String racers are an effective way to demonstrate Newton's third law. This set of instructions gives plans for two types of string racers: the simple balloon racers that are quick and easy to build and a rubber band–powered propeller design that can race down a string 80 ft. or more. Both of these pieces can be built and demonstrated simply as an example of action-reaction, or they can be incorporated in a project-building assignment for students to test a variety of design objectives.

Student Skills

Observation. The string racer and the balloon will move forward as a result of air pushed backward.

Design and Construct. Students can build a copy of the string racer with the plans provided, or they can design their own version and try to improve its performance. Other variations of the propeller and tube can be adapted for use on land or water.

Application. Students can look for other examples of propulsion or forward movement as a result of air being pushed backward.

Related Concepts or Processes

Potential energy	Friction
Kinetic energy	Newton's second law
Inertia	Newton's third law
Simple machines	Speed
Design	Testing of hypothesis
Trial and error	Construction techniques

Prior Knowledge

You can use these pieces for early elementary students if you use them as a demonstration. Younger students should be able to define motion. Prior introduction to the application of force as a push or a pull is helpful. They should understand that forces are needed to

---

SAFETY NOTE

The moving propeller can hurt fingers or cut skin if it hits someone while it is spinning. Students must be careful not to let the spinning propeller hit their fingers upon release and must not try to catch the racer before it stops.

The string racer can move quite quickly down the string, possibly jumping off the holder along the way. Instruct students to stand back from the moving string racers.

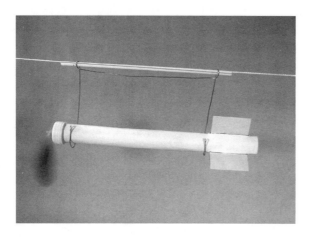
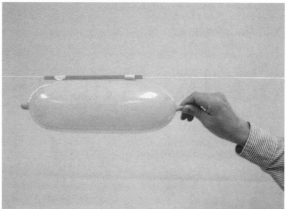

make things move and that friction slows objects down and eventually stops them. Older students should have some basic understanding about motion and how we can measure it. They may have been introduced to some ideas about storing energy versus energy being used. These pieces also can be used to demonstrate Newton's laws of motion.

Predemonstration Discussion

Students can review the ideas of how forces are applied and energy is needed to make objects move through numerous examples. You should demonstrate wind-up toys, lift a ball up and then allow it to roll down a table, and show the balloon racer before introducing the string racer. A number of questions are applicable; use those that apply to the age of your students. Some questions for discussion include the following:

- What is needed to make an object move?
- How can we add energy to make things move?
- What are some ways that toys or objects can store energy?
- What are some ideas that affect the distance that an object will move?
- Why do moving objects slow down?

Suggestions for Presentation

Thread a long string through a drinking straw and then tie it so it extends from one end of the room to the other. A successful sequence might start with a demonstration of a deflated balloon and a discussion of how energy can be added to the balloon. After energy is added, students can predict how the balloon will move if it is released. If you have other shapes of balloons, discuss what flight characteristic each might have.

The discussion about the untethered balloon should lead into attaching the balloon to the string. This allows for a more efficient movement. The string

helps stabilize the movement of the balloon and directs the air backward. The balloon is then propelled forward, usually at a higher rate of speed than in the untethered mode, because less energy is lost to extraneous movements.

The propeller-driven string racer can follow the balloon racer. Attach the wire holder to the string. It is easier to wind up this string racer first and then attach it to the holder. Try not to pull down on the string racer as the propeller is released because that will make it bounce up and waste a lot of energy.

Interactive questions about balloon energy could include the following:

- Does a deflated balloon have any potential energy?
- How can we put energy into it?
- What happens if I let the inflated balloon go?
- Why does it not travel in a straight path?
- How can we get the balloon to travel in a straight path?

Interactive questions about the balloon racer could include the following:

- How does the flight characteristic change when the balloon is tethered to the string?
- Compare the speed of the tethered and the untethered balloons.
- Why does the tethered balloon move faster?
- If the balloon goes forward, what must the particles of air do?
- How would the amount of air put into the balloon affect the speed and distance that it travels?
- How can we calculate the speed of this racer?

Interactive questions about the propeller-driven racer could include the following:

- What are the similarities between the balloon racer and the propeller racer?
- What are the differences between the balloon racer and the propeller racer?
- How does this vehicle move?
- How does a propeller push air backward?
- What are some variables that might change the speed and distance that the racer travels?
- How can we calculate the speed of this vehicle?

Postdemonstration Activities and Discussion

The propeller-driven string racer was originally designed as a building project for elementary students. A pattern was printed on heavy stock paper by using a copying machine. The bottle caps had holes added to them, and students assembled the propeller system. It is a simple project that fourth- and fifth-graders can build with some assistance from an adult. When used as a demonstration for middle school students, this propeller-driven racer generated the same enthusiasm it did among the younger children. Knowing that the purpose of the fins is to keep the string racer from turning as it travels from one end to the other, middle school students found it interesting to use larger fins and adjust them so that the string racer would do two or three loops as it traveled from one end to the other.

Propeller-driven string racers can also be the basis for a design project. Give students a variety of materials, and challenge them to construct a device that will travel down the string as quickly as possible. You can show the string racer as an example of how it can be done, although this usually results in students' copying it and may limit the designs students will try.

Figure 1 is a typical example of a student-designed string racer. In this case, students were not shown any examples before they were given the assignment.

Students tried using the propellers to push or pull the string racer. Both ways can be successful. Most students stuck with a single propeller; some tried using two propellers, and a few tried three. Usually, the simplest designs are the most effective.

Suggested Materials

Purchased propellers or propellers made from tongue depressors, soda bottle caps, soda bottles, cardboard, paper, mat board, 18-gauge steel wire, foam meat trays, paper clips, 8 in. long rubber bands, balsa wood, glue gun, tape, wheels, K'nex, LEGOs.

The string racers can demonstrate several concepts of force and motion. For example, they show potential energy changing into kinetic energy. The balloon racer

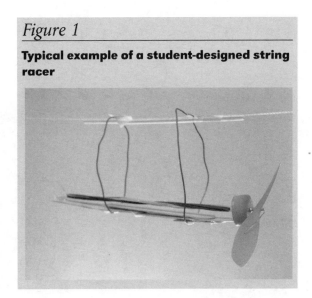

Figure 1

Typical example of a student-designed string racer

stores energy (potential) by stretching the latex fabric. This energy changes into kinetic energy by pushing the gases out of the end. The string racer stores energy by twisting the rubber band and uses the propeller to pull the device through the air as it pushes air backward.

String racers can demonstrate Newton's laws of motion. The devices continue traveling forward even after the balloon empties or the rubber band unwinds. This is an example of inertia. Both of these pieces slow down and eventually stop due to friction between the string and the straw. Newton's second law explains the relationship between force, mass, and acceleration. The movement of the racers can demonstrate how an increase of mass can decrease the acceleration. You can also demonstrate Newton's second law by changing the force. The amount of force is determined by how many rubber bands are turning the propeller, so changing the number of rubber bands affects the acceleration. Newton's third law talks about action and reaction. Both of these devices go forward by applying a force backward. In either case, students can feel the air particles being moved backward.

Additional Activities

Have students keep a journal to trace their progress through this problem-solving activity, designing sketches that can be part of the journal and including ideas that worked and those that did not.

Another form of transportation for the propeller-powered tube can be some type of wheeled car. The base supports are carved from foam insulation material. The wheels and axles fit through a plastic drinking straw. The straw is glued into place using a hot glue gun. Rubber bands are used to attach the propeller tube to the supports.

SAFETY NOTE

The moving propeller can hurt fingers or cut skin if it hits someone while it is spinning. Students need to be careful not to let the spinning propeller hit their fingers or skin.

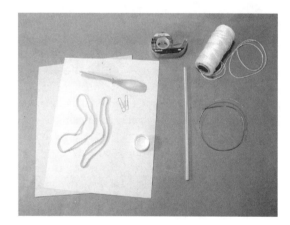

DIRECTIONS FOR ASSEMBLY

Propellers

Materials
- 2 sheets of heavy stock paper
- 1 soda bottle cap, any size
- 2 small paper clips
- 1 small plastic bead
- 1 plastic drinking straw
- 3 rubber bands
- 24 in. of thin gauge metal wire
- 1 plastic propeller or tongue depressor for homemade propeller
- Plastic or masking tape
- 50-plus ft. of string
- Scissors
- Large dowel rod
- Ruler
- Sharp-pointed object
- Pliers
- Sandpaper or nail file

The string racer and variations are all based on the same design using the tube and propeller as the power supply source.

1. To make the tube for the body, start by cutting three slits into the bottom corner of the heavy stock paper. The slits are needed for the fins of the tail section to fit through. Cut the slits 2 in. long and 1 in. apart. Start them ½ in. from the right side edge and ½ in. from the bottom edge. Use a pen to draw thick lines onto the stock paper.

2. You can cut the slits using the edge of the scissors as a cutting edge.

3. After cutting the slits for the fins, roll the paper into a cylinder shape a few times to help retain this shape. Use a large dowel rod as a guide for the rolling process. Make sure the slits appear on the outermost layer of this tube instead of on the inside.

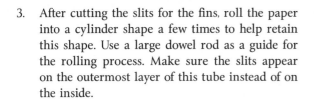
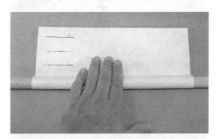

4. The tube will eventually fit snugly in a soda bottle cap. Use the bottle cap to determine the correct size.

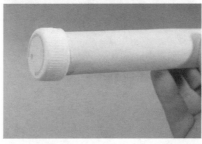

5. Make the fins for the tail section from another piece of heavy stock paper. Use a ruler to measure and then draw a pattern that is 2 in. wide and 10 in. long. Use a ruler to measure and then mark off 1 in. blocks for the folding lines. Draw nine lines across the 2 in. width of the tail section.

6. Cut out the 2 in. × 10 in. section with scissors.

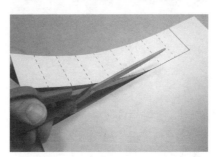

7. This pattern is for folding the tail section. Each line indicates where to fold the heavy stock paper.

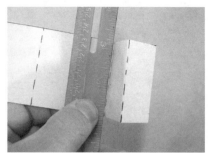

8. Use a straight edge in the folding process to help ensure the folded lines match the drawn lines. The folds must be accurate to ensure that they match up with the slits cut in the body tube.

9. The creases in the paper must be well defined. Lay the fold on a hard surface and press and slide the ruler against the fold. Do this with each fold.

10. Fold the tailpieces tightly together. Each tail fin is made from two pieces of the paper folded together. You can cut these fins to resemble a rocket-tail assembly after they are inserted in the tube body.

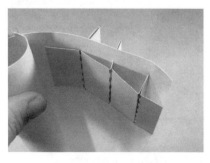

11. Unroll enough of the body tube so that the folded tabs for the tail section can be inserted into the slots cut into the body tube.

12. Roll the paper and fins back into the cylinder shape again. The fins should be evenly spaced in the back end of the cylinder.

The fins help stabilize the string racer as it moves on the string. They can be adjusted to make the string racer do loops as it travels down the string.

13. The tube diameter should be approximately the same as the inside diameter of the bottle cap. Check the tube diameter at the front and back ends of the tube using the bottle cap.

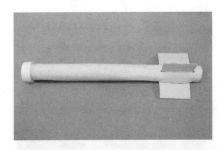

14. Apply a few pieces of tape in the tail section area and along the tube body to keep the tube from unrolling.

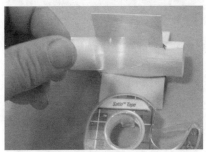

15. To assemble the propeller, start by removing the inner liner from a soda bottle cap and then use a pointed object such as an ice pick or compass to poke a hole through the center. The hole should be just slightly larger than the diameter of the paper clip.

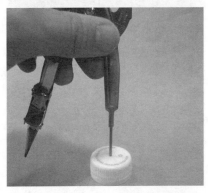

16. Make the shaft for the propeller from a paper clip. Unbend two of the three bends. Leave the smallest of the three bends in place. This shaft should resemble a *J*. If the paper clip is too long, cut it using the cutting edge of pliers.

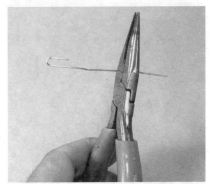

17. For a smoother operation, the propeller should turn in a straight path without any wobbles. The area where the shaft goes through the propeller is rounded. Sand the rounded tip of the propeller shaft flat, using a fine-grit sandpaper or nail file.

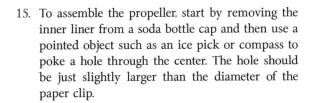

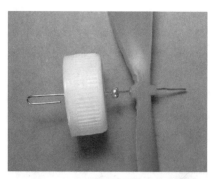

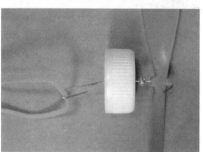

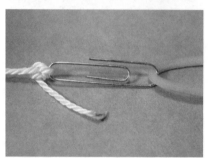

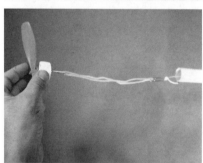

18. Begin the assembly of the propeller by pushing the straightened end of the paper clip through the hole in the soda cap. The hooked end of the paper clip is toward the inside of the bottle cap. Now add a small bead with flat ends to the paper clip shaft, followed by the propeller. The bead will act as a washer between the cap and propeller and allow it to turn more efficiently.

19. Use the pliers to bend over ¼ in. of the paper clip to catch the propeller.

    The speed of the racer is dependent on the number of rubber bands used to propel it; three or four usually work very nicely. Hook the rubber bands onto the *J* portion of the paper-clip shaft. Bend the end of the paper clip over the rubber bands to hold them in place.

20. Hook the other end of the rubber band onto another paper clip. Tie a short string onto this paper clip to ease the fitting of the rubber bands through the body tube.

21. To assemble the propeller onto the body, hold the tube vertically and dangle the string into the tube. After the string comes through the opposite end of the racer body, pull on the string to stretch the rubber bands.

22. To attach the rubber bands to the racer body, hook one of the bends of the paper clip onto the tail section of the tube body.

23. The body of the racer is completed. Wind it up and test it a few times to see how it spins. The propeller should spin without wobbling. If the propeller wobbles, then the paper clip shaft needs some minor adjustments. Do not overwind the rubber bands, or they will break inside the tube.

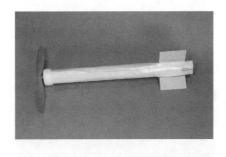

The final assembly of the string racer requires that the body be attached to a string that is tied across the room.

24. A simple method for attaching the racer to the string uses a drinking straw and 24 in. of metal wire. The wire can be bent to hold the racer body to the straw and allow easy attachment or removal of the racer for winding or repairs.

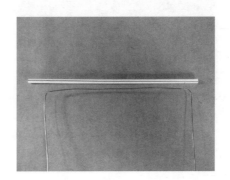

Start by threading the string through the straw and then tie the string onto two supports.

Measure and cut off about 24 in. of the wire. The middle 8 in. of the wire remain straight against the straw. Bend the outer sections 90° to form a large *U* section.

25. Bend over the top ½ in. of the 8 in. wire section to clip onto the straw. Use a dowel rod or pencil to form the bend in each corner of the wire. Don't make the bends too tight, or they will pinch the string inside the straw when it is attached.

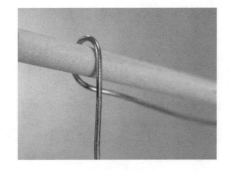

26. Make the second set of bends in the wire hanger about 6 in. down from the straw. Wrap a 1 in. circular bend in the wire around the racer body to hold it in place. The bends must hold the racer enough beneath the straw so that the propeller does not come in contact with the string. Wrap the wire around the racer body for about three-quarters of the distance around. Cut off the excess wire using wire-cutting pliers. The last ¼ in. of the wire should be folded back on itself to form a very small loop that makes a blunt edge.

27. The racer body is now ready for a practice run. Wind the rubber bands up by turning the propeller. Place the racer back onto the wire supports while holding the propeller and then let it go.

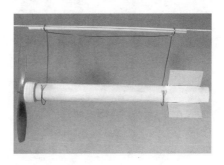

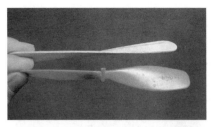

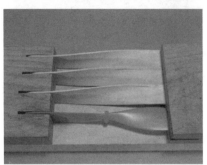

Alternate propellers

An alternative to purchased propellers is having students make their own. Several designs can be built and tested. Typical purchased propellers have the blades offset by 90°.

1. A tongue depressor can be formed into the same shape as a purchased propeller by twisting it 90° from one end to the other.

2. If you soak the tongue depressor in water for about an hour, it will become soft and pliable and can be easily bent. A simple form for holding the tongue depressors has four horizontal and four vertical slits about 4 in. apart. Place the wet tongue depressors in this form until they dry. After they dry, they will retain the 90° twist. Then drill a hole in the center for the propeller axle.

Balloons

Balloons

Materials
- 1 straw
- 1 long thin balloon
- Masking tape
- 50-plus ft. of string

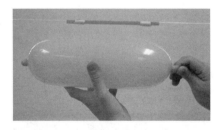

1. Thread a string through a drinking straw and tie it to two supports about 20 to 40 ft. apart.

    Wrap a short piece of masking tape with the sticky side out on the front edge of the straw. Wrap another piece of tape with the sticky side out on the other end of the straw.

    Do not wrap the tape too tightly around the straw. Success depends on whether the tape can slide back and forth on the straw with very little effort.

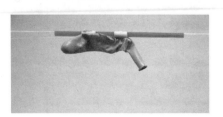

3. To operate this racer, blow up the balloon to its maximum size. While holding the balloon, lift it to the straw and press the center of the balloon onto the sticky sides of the tape.

    Release the balloon when you are ready. The balloon should race down the length of the string.

4. A successful trial should have the balloon still on the straw with both tapes moving toward the center of the straw.

This chapter first appeared in If You Build It, They Will Learn: 17 Devices for Demonstrating Physical Science *(2006), by Bruce Yeany.*

Chapter 4

Paper Towers

Building Students' Understanding of Technological Design

by James Minogue and Todd Guentensberger

A seemingly simple question was posed to a group of undergraduate education majors during a science teaching methods course: What do you think the National Science Education Standards are referring to when they talk about *science and technology*? Immediately, the students, having previously completed an assignment in which they examined the NSES Content Standards, suggested computers, the internet, probeware—and then there was a protracted silence. Our suspicion is that the above-described scenario is not unique to these particular students. In fact, we suggest that this somewhat myopic notion of technology is a widely held misconception that may be part of the reason why it has proved difficult for educators to develop and assess the specific abilities and concepts that underlie technological design.

The above question was purposeful, serving as a way to introduce the activity for the day: Paper Towers,

an exercise that we both used when we taught middle school science (*www.ciblearning.org/pdf/Exercise.PaperTowers.pdf*). One set of ideas at the core of the NSES Science and Technology Standards is that of engaging middle school students in activities that help them develop their understanding of technological design. More precisely, students should be able to identify appropriate problems for technological design, design a solution or product, implement a proposed design, evaluate completed technological designs or products, and communicate the process of technological design (NRC 1996). This article describes a simple yet effective classroom activity that can help students achieve many of these learning goals.

BUILDING PAPER TOWERS AND UNDERSTANDING

Engaging students in a paper tower construction activity starts with a challenge: Build a tower as tall as possible that will resist being blown over by the teacher from one arm's length away. Students may use only the materials that are listed (two sheets of newsprint, 25 cm of transparent tape, scissors, and a ruler). The person who constructs the tallest standing tower is the winner. Next, students are given a set of design constraints, which include working individually, only

Further Reading

- "The Tower Challenge," from *Activities Linking Science With Math, 5–8* (2009)

 PreK-5

Building a Paper Tower

Challenge

Build a tower as tall as possible that will resist being blown over by a hair dryer set on low at a distance of 1 m. You may use only the materials that are listed below. The person who constructs the tallest standing tower will be the winner.

Materials

- Two sheets of newsprint
- 25 cm of transparent tape
- Scissors
- Ruler

Constraints

- Work individually to construct your own tower.
- Tape can be used only to attach paper to paper.
- Tape may not be used to secure the tower to the floor or any other object.
- Paper can be measured, cut, torn, and folded in any way necessary.
- You have approximately 30 minutes to construct your tower.

Procedures (for after your tower is completed)

1. Draw a detailed sketch of your completed tower.
2. Measure your completed tower from its base to its highest point. Record your measurement in centimeters:
   My tower is ____ cm tall.
3. As you watch your classmates' towers being tested, record your observations. Focus on the features or characteristics that made the tower a success or failure.
4. Did your tower continue to stand after the teacher tried to blow it over?
   - If yes, explain why you believe your tower was successful. How did your tower withstand the wind?
   - If no, explain why you believe your tower was unsuccessful. Explain how you would improve your tower if you were given a second chance.
5. Based on our class discussion, create a list of Successful Engineering Principles for Paper Tower Design.

using the tape to secure paper to paper, and having only 30 minutes to plan and build the towers. Students are prompted to think about a plan for their towers, which may include a rough sketch, before beginning construction. Once construction begins, the teacher's primary role becomes one of an observer, but you may chose to discuss designs and ideas with individual students as you circulate.

After students complete their towers, they work on the first two items under the procedures section of the activity sheet. These items ask students to sketch and measure the height of their creations, allowing them to practice important science skills. Each tower is then subjected to a wind test, while students make careful observations. Students are asked to focus on the characteristics of each tower that led to its passing or failing the test and to record their observations.

The 30-minute construction time should leave enough class time to test each tower. Regarding the testing method, we have always filled our lungs with air and subjected each tower to a strong and steady stream of air. The teacher might want to exercise tighter control over the wind force and avoid the possible spread of germs by using a fan or a hair dryer. As one might imagine, the tower designs are usually varied, as shown in Figure 1, and most are not successful. That is, they often fall easily when subjected to the wind test and many are even unable to stand on their own.

Because of time constraints, you may need to assign procedures 4 and 5 as homework. This will give students ample time to reflect on their designs, synthesize ideas based on their observations, and devise ways to improve their designs. During the next class period, engage students in a discussion that draws on their observations during the testing process. Ultimately, this interactive dialogue results in a list of critical features or attributes that the few successful towers possess (or that unsuccessful ones lack). Students compile these attributes into a list of Successful Engineering Principles for Paper Tower Design, which may include a strong, wide base; a low center of gravity; a small surface area in contact with the wind; and a support system with tensile strength.

A multitude of opportunities exist to tie these ideas into science concepts such as gravity and the relationship between structure and function in both naturally occurring and designed objects. For example, rolling the newsprint diagonally into hollow tubes affords the structure a lot of strength, which can be likened to carbon nanotubes or the cell's microtubules. Following this content-rich discussion, students create a list of successful design principles.

THE DESIGN LOOP AND ITS BENEFITS

The next time the students come to class, have them redesign and rebuild a paper tower using the same activity sheet and protocol used previously. The improvements of the redesigns are amazing, as represented by the images in Figure 2. Not only are the redesigned towers taller and stronger, many are able to withstand the wind test.

We feel that allowing students to redesign and rebuild their towers has numerous benefits, both cognitive and affective. First, it allows students to apply their knowledge of the basic principles of tower design. Additionally, it engages them in key aspects of a design loop that requires students to choose a solution, build a prototype, test it, evaluate it as compared to their original designs, and use this information to inform a redesign.

Having an opportunity to learn from their mistakes promotes a metacognitive approach to this activity. Students identify and think more deeply about each step they take to overcome previous design problems. During the rebuilding phase, you may choose to have students work in small groups to promote teamwork and emphasize that real scientists rarely work alone.

CONCLUSION

This activity offers a challenging and engaging way to introduce students to the central tenets of technological design. This type of lesson helps students build a more complete understanding of the meaning of *science and technology* as put forward by the National Science Education Standards—and students have fun in the process.

Reference

National Research Council (NRC). 1996. *National science education standards*. Washington, DC: National Academies Press.

Figure 1

Representative images of student towers

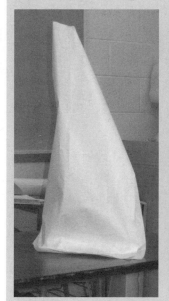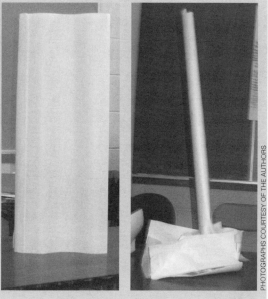

PHOTOGRAPHS COURTESY OF THE AUTHORS

Figure 2

Sample redesigned towers

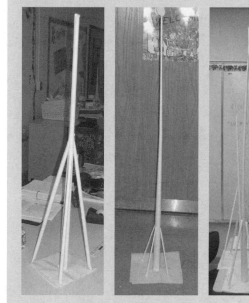

Resource

Center for Inquiry-Based Learning (CIBL)—*www.ciblearning.org*

This article first appeared in the November 2006 issue of Science Scope.

PART

2

Teacher-Created Constructions and Repurposed Materials

Chapter 5

Discovery Bottles

A Unique, Inexpensive Science Tool for the K–2 Classroom

by Sandy Watson

Discover discovery bottles! These wide-mouth plastic containers of any size filled with objects of different kinds can be terrific tools for science explorations and a great way to cultivate science minds in a K–2 classroom. I've found them to be a useful, inexpensive, and engaging way to help students develop skills in making predictions, observations, and comparisons.

The bottles work well for use in learning centers or stations as starting points to get students excited to learn about a topic. They are also useful for independent observation and exploration of science concepts in a concrete way. Depending on the concept explored, some bottles may need to be sealed with a glue gun (if you don't want students to have contact with the contents), while others may be left unsealed so their contents can be accessed. The possibilities are limitless. Here are few of my favorites.

MAGNETIC THINGS

Objective

Students will explore magnetism by attempting to attract objects of different materials with a magnet. Students will predict which materials will be attracted to a magnet and which will not. This is a standard lesson but is enhanced with the use of discovery bottles

because students can investigate independently or with a partner with simple materials safely housed within a sealed bottle.

Materials

- Sealed bottle one-third full of sand and filled with magnetic and nonmagnetic objects such as metal paper clips, buttons, pennies, and beans. Make sure that some items that are magnetic are different colors. Include some items that are metal but not magnetic (e.g., items made of copper, aluminum, or other nonmagnetic metals)
- Worksheet (Figure 1, p. 28)
- Magnetic wand (purchased in dollar stores or most toy stores)

Exploration

Give students a magnetism discovery bottle to observe. Ask, "Which items in the bottle will be able to be moved by the magnetic wand and which will not?" Instruct students to underline on the worksheet the objects they think will be attracted to the magnetic wand. Have students provide evidence for their predictions and share their reasons why they think a particular object would be attracted to the wand or not. Then have students test their predictions, moving the wand up and down the sides of the bottle. Afterward have

Figure 1

Magnetism discovery bottle worksheet
(Sample answer from students in italics)

PREDICT which of the following items will be attracted to the magnet by underlining them.

Run the magnetic wand up and down the sides of the bottle. Circle the items on the worksheet that you were able to move with the magnetic wand. (Object number eight is a necklace/chain.)

COMPARE what actually happened to what you thought would happen. Based on your observations, what do you think makes a material or object attracted to a magnet?
It has to be made of something metal.

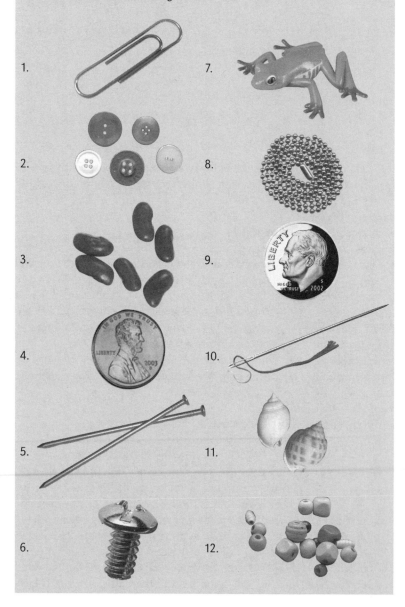

1.

2.

3.

4.

5.

6.

7.

8.

9.

10.

11.

12.

students circle on the worksheet the items they were able to move with the magnetic wand (magnetic objects) and compare these answers to their predictions.

Discussion

Students answer the question "What makes a material or object attracted to a magnet?" Most students believe that the materials or objects must be made of metal. Students investigate this idea by using the magnetic wand to attract the objects and then report what was attracted and what was not. They then look at all the items they circled (magnetic) and compare them to one another and to the nonmagnetic (not circled) objects. To guide students, ask, "What do all the magnetic objects seem to have that are alike? Did you make predictions that were incorrect? Which ones did you get wrong? Why do you think these items were magnetic? What do you now know about magnetic materials that show these items to be nonmagnetic? Compare the magnetic and nonmagnetic items and tell how they are different."

Students should recognize that not all metals are magnetic. If the students still say that all metal things are magnetic, ask, "Did the magnet attract all of the items made of metal? Point out that the magnet only attracted *some* of the metallic items—the magnet was not attracted to aluminum foil, for example.

Assessment

The instructor assesses students' progress via group discussions, student performance on the handout, or individual conversations with students.

Objective

Students will explore static electricity as they rub the static electricity discovery bottle (a bottle filled with tissue paper and other light items) on their hair or shirts. Students will also learn under what conditions static electricity might be increased or reduced (using friction, weather conditions, etc.).

Materials

- Bottle filled with various items, such as small pieces of foam from a foam coffee cup, small pieces of paper, or beans
- Worksheet (Figure 2)

Exploration

Distribute the static electricity discovery bottle and have students observe the contents and try to identify what is inside. Students typically describe the contents as little pieces of paper, beans, and so on, but they may not be able to figure out what the foam pieces are or how to define them. The teacher might ask if the unknown pieces (foam) appear hard or soft and where they might have seen such material before. If the students still do not know what the foam pieces are, a Styrofoam cup should be produced.

Next, on their worksheets, have students record what they think will happen to each object in the bottle if they rub the bottle on their heads or shirts. Student answers commonly include *the beans will make noise and not stick to the side of the bottle, the pieces of paper will "jump around,"* and *the foam and pieces of paper will stick to my head through the bottle.* Ask why they believe that—students may have some previous experience with static electricity.

Now have students rub the bottle on their heads or shirts or against the carpet and observe what happens. Students observe that the beans just make noise and fall back to the bottom of the bottle when it is set back down; the pieces of paper may cling to the sides of the bottle; and the foam pieces may also cling to the walls of the bottle. If they keep playing with the bottle after rubbing it on their heads and shirts, they may see that the charged materials (pieces of paper and foam) may jump around as they touch the sides of the bottle.

Have students compare what they actually observed with what they predicted would happen to each object.

Figure 2

Static electricity discovery bottle worksheet
(Sample answers from students in italics)

OBSERVE the contents of the bottle.
Draw or write what the contents of the bottle look like and seem to be made of.
It looks like little pieces of plastic and paper.

PREDICT what will happen to the contents of the bottle if you rub the bottle on your head or shirt.
The little pieces will fly all around inside the bottle.
Rub the bottle on your head or shirt and observe what happens. Draw or describe what happened.
The little pieces stuck to the sides of the bottle.

COMPARE what actually happened to what you thought would happen.
I thought they would be jumping around but they stuck to the sides.

EXPLAIN why you think this happened.
The charge on the bottle is different, and the paper pieces will stick because they have a different charge.

Students are often surprised to see the pieces of paper and foam clinging to the bottle's sides.

Discussion

Ask students if they have ever been "shocked" after shuffling across carpet and then touching a doorknob. Or if they have ever rubbed a balloon against their hair and stuck it to a wall. Tell them that what they experienced was *static electricity.* Explain that the material that the bottle is rubbed against (hair, shirt, carpet) has become positively charged. Like charges repel. If students have experiences with magnets, compare this to a magnet, and the positive and negative sides of a magnet. Unlike charges attract. The now negatively charged items in the bottle are attracted to the now positively charged materials the bottle was rubbed against, allowing those items to become attracted to and cling to the positively charged material.

Assessment

Student knowledge may be assessed via group discussion, question-and-answer sessions, evaluation of the student worksheet, or individual conversations and demonstrations.

chapter 5

Figure 3

Sink or float discovery bottle worksheet
(Sample answers from students in italics)

Obtain the sink or float discovery bottle along with a bag of objects from the science learning center. Below are pictures of each item in the bag:

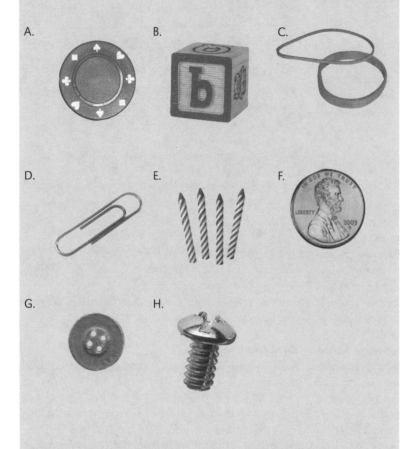

A.   B.   C.

D.   E.   F.

G.   H.

PREDICT which objects will float and which will sink when dropped into the bottle by underlining the pictures of the objects you think will sink.

Drop each object into the bottle one at a time. If the object floats, circle it. If it sinks, do not circle the item.

COMPARE your predictions to what actually happened.
I got most of them right except I thought the button would float and it didn't.

The penny and the block of wood weigh the same. But the penny sank and the wood floated. Can we say that heavy things sink?
No, some things weighed less and sank. Some things weighed more and floated.

Can you think of something that is heavy but floats?
Boats, people, a dock

Objective
Using a magnifying glass, students will observe items in a discovery bottle and predict which will sink and which will float.

Materials
- Bottle filled with various items, such as a poker chip, a small block of wood, a rubber band, a penny, a paper clip, a birthday candle, a button, and a bolt. Include items of different weights but similar volume and shape so students do not confirm a common misconception that heavy things sink.
- Worksheet (Figure 3)
- Water source
- Towel
- Magnifying glass

Exploration
Distribute a sink or float discovery bottle, a magnifying glass, and a towel to students. Have students open the discovery bottle and pour the items out onto the table. Give each student a sink or float worksheet and have them predict which items they think will sink in water by underlining their pictures on the worksheet. Next partially fill the bottle with water. Then have students drop each item, one at a time, into the bottle of water and circle the item on the worksheet if it floats. Finally, they compare their predictions with what actually happened.

Discussion
Ask students what they already know about sinking and floating. What kinds of objects tend to sink? Answers may include heavy things, things that weigh more, or things that are not light. Ask students to give you examples of objects that

chapter 5

they think would sink. Ask what they all have in common. Then ask the students to identify the types of objects that tend to float. Answers might include things that are not heavy, things that can swim, or things that are flat. Ask students to identify specific items that they have seen floating (ball, leaf, boat). Try to look at the examples they have provided, and with the students determine what characteristics they have in common. Answers might be that they have air in them and that they have flat shapes. Then go over what happened in their activity. Identify the items that floated and the ones that sank. Ask how many of their predictions were correct. Ask if the predictions got better or worse as they went along, or did the predictions stay the same. Then ask the students to put all the objects that sunk into one pile and all that floated into another. Ask them to look at the items that sank and describe them. See if the students can identify what characteristics they have in common. Do the same with the float pile. Did everyone get the same results? If any were different, ask the students to try it again in front of the class.

Building on this discussion, students might identify similarities and differences in the materials that make up the objects that float and those that sink. They may state that objects made of wood usually float and those made of metal usually sink. They may again state that the weight of an object determines whether it sinks or floats. If students still believe heavier things sink, pick two objects that have similar volumes but different weights so that one sinks and one floats. Discuss what happens. Did the heavier one sink? No, they weighed the same? At this age students should not be introduced to density but they should have experiences that challenge the idea of heavy things sinking.

Connecting to the Standards

This article relates to the following *National Science Education Standards* (NRC 1996).

Content Standards
Standard A: Science as Inquiry
• Abilities necessary to do scientific inquiry (K–4)

Assessment

Student assessment for this activity may take place through the evaluation of the worksheet or what is said in group discussions or singular conversations. You might bring in a new set of objects (such as pool toys) and ask whether they would sink or float in a swimming pool and then try each in a tub of water. Toys to include in this assessment might be a Frisbee, heavy diving egg, or toy boat.

These are just a few discovery bottles ideas for use in K–2 classrooms—the possibilities are endless. Discovery bottles are inexpensive, quick to assemble, and an excellent way to provide students with practice in developing science-process skills, such as observing, measuring, predicting, and so on. I highly encourage you to mix up a few bottles of discovery yourself. They're worth it!

Reference

National Research Council (NRC). 1996. *National science education standards*. Washington, DC: National Academies Press.

This article first appeared in the Summer 2008 issue of Science and Children.

Materials Repurposed

Find a Wealth of Free Resources at Your Local Recycling Center

by Orvil L. White and J. Scott Townsend

Few teachers find themselves with the support to purchase all the materials they ideally need to supply their classrooms. Buying one or two simple, ready-made items can put a serious strain on anyone's budget. However, materials for science in the classroom need not be prefabricated or expensive. By looking at the function and purpose of any piece of equipment, a creative teacher can find a suitable replacement for many premade science materials, sometimes from the most unlikely places. This is not to say we advocate the potentially hazardous practice known in some circles as "Dumpster diving," but with proper caution and common sense—like partnering with your county's local recycling center—you can find some terrific, serviceable materials among what others have deemed "trash."

Our local recycling center offered a community outreach program called "Materials for the Arts," in which public and homeschool teachers in the county had access to a wealth of materials salvaged from or donated to the recycling center. The center dedicated two rooms at the facility to the program, which stored objects such as clean, sanitized containers of all sizes, including plastic and glass bottles, coffee cans, potato chip cans, baby food jars, and cereal boxes and oatmeal containers; as well as cardboard tubes, carpet squares, compact discs, plastic trays, corkboard, bubble wrap, and other things. All of the materials were required to be completely cleaned with, depending on the material, either antibacterial soap or Lysol spray, before being accepted for donation. Use salvaged materials only if they have been thoroughly sanitized.

If there is no such program in your area, you might consider starting one at your school. Local recycling centers are often looking for outreach opportunities. When we conducted a presentation at our state science teachers' conference a few years ago, several outreach personnel from various state recycling centers approached us for ideas about how they could perform the same outreach services to teachers and the community.

Sometimes we visited the recycling center with specific material needs in mind. Other times, though, we simply explored the rooms to see what ideas were sparked by the materials at hand. Of course, not every item at the recycling center can be repurposed into a useful tool for the science classroom, but here we share a few of our favorites.

TIMER

Teachers can make a classroom timer from two plastic drink bottles with caps and an old 35mm film canister with the bottom removed. Glue the bottle tops together inside the film canister with the tops touching. Trim off the excess canister with kitchen shears, and using an electric drill with a 3/16 bit, drill a small hole through both caps. Place approximately 800 g (for a five-minute timer) of sand, salt, or sugar into one bottle and screw the cap in place. Next, invert the second bottle and screw it into its cap. Test it out and adjust the amount of material as necessary for the time desired. This timer can now be used in a variety of ways, including as a guided-inquiry activity model for students to create other timers of various durations. The timer itself can be used to time speakers, give "time remaining" for a quiz or other assessment, and allow students to better understand the ways in which time has been historically measured.

Using different models of timers, students can investigate questions such as the following: Is there a difference in using sand, sugar, or salt? How does the size (diameter) of the hole in the caps affect the rate at which the materials flow? How does the particle size affect the time it takes for the grains to go through the opening? If using a material other than sand, does the flow rate change over time, and if so, is it faster or slower? Upper elementary or middle school students could also create a graph to show the relationship between mass (in grams) of granular material versus the amount of time it takes for the material to completely travel through the container. This would ultimately allow students to predict how much material they would need to insert for resulting amounts of time in the timer.

SEDIMENT TUBES

Any plastic tube or bottle can be used to show the sedimentation of materials through a water column. We used plastic tubes made from a fluorescent lightbulb cover, a plastic sheath you can buy to cover the bulb, which we cut to size using kitchen shears. The covers are available at most home center outlets. Pour sand or soil into the bottle and fill with water. Know the source of the soil to avoid contaminants. Make sure children wash hands thoroughly after handling soil. Shake and allow the material to settle. The students can observe the soil settling into layers based on the density of the materials contained in the soil.

The sediment tube allows students to model and observe the process of deposition of materials in the natural environment. This process is the prelude to the formation of sedimentary rocks in the Earth's crust. The process of deposition of materials can be used to show how, over geologic time, rocks with differing colors of strata are formed. Students can also use this

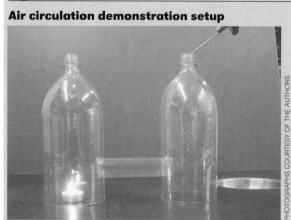

Figure 1

Air circulation demonstration setup

PHOTOGRAPHS COURTESY OF THE AUTHORS

method to separate different soils into parts according to grain size and, by measuring the thickness of the layers, calculate percentage of each part—thereby adding a link to mathematics standards. Also, sediment contamination of streams and rivers is an issue in environmental science that can be better visualized and understood once the students can see how soil breaks down and is deposited when mixed with moving water.

DEMONSTRATING CIRCULATION

Recycled materials can also be repurposed for a teacher demonstration exploring air circulation behavior. Remove the bottoms of two 2 L plastic bottles and connect them with a plastic tube, actually a fluorescent lightbulb cover that was cut to size with kitchen shears. Place a small candle under one bottle and hold a lighted stick of incense over the other (see Figure 1). Do this as a teacher demonstration only. Use a tea candle and keep matches out of reach of children. The heated air should rise from the top of the bottle and produce a low-pressure area, drawing the air from the higher-pressure area of the other bottle. This will cause the smoke from the incense to flow down, flow across, and rise with the heated air out of the top, demonstrating the process of air flow in weather systems.

The demonstration models the movement of air in the environment. Air that has been warmed rises, and cold or cooler air moves in to take the place of the warm air. When used as part of a unit on weather, this demonstration enables students to see a process that is generally unobservable and helps explain the shifting wind patterns they can feel. It is useful in exploring sea, land, mountain, and valley breezes, as well as the displacement of warm air when a cold front moves across the landscape. Additionally, this is a good model of how other fluids react when heated. Ocean currents and the movement in a pot of boiling water are other concepts linked to thermal circulation.

GRADUATED CYLINDER AND SCOOPS

You can make a graduated cylinder by measuring a known volume of water into an old plastic bottle, with the label and bottom removed, and marking the measurement with a permanent marker. A clear 1 L water bottle works best for larger volumes, and any smaller straight-sided bottle will work with lesser amounts.

Cutting the bottom of a 1 L bottle will create a scoop that is easy both to use and to pour material from. Scoops can also be made from old salad-dressing bottles cut along the bottom and side. The caps should be glued in place to prevent accidental spills.

FUNNEL AND CUP

A simple funnel can be made by cutting the top off a 2 L bottle and inverting it so the small opening is at the bottom. Aside from their usual use, funnels can be used as part of an inquiry challenging students to design the "most effective" water filtration device. Give students a choice of materials (e.g., coffee filters, paper towels, sand, activated charcoal, shelf liner [the puffy, nonslip type], gravel, cotton balls, sponges, and so on) to design a three-layered water filter within the plastic funnel to remove a small scoop of potting soil from a water sample. Follow all safety rules when working with soil. Know the source of the soil to avoid contaminants and wash hands thoroughly with soap and water after working with soil. The goal for the student groups is to design a filtration system that will result in "clear" water being produced in a timely manner. This activity can be used as a stand-alone inquiry or as part of a larger unit on soils/Earth materials, water quality, or mixtures and solutions and the separation of their component parts.

MYSTERY BOXES

Mystery boxes are a favorite tool for teachers to introduce the meaning of observation and inference and various aspects of the nature of science. Often they are made by purchasing small cardboard boxes from the local jewelry store and placing common classroom or household items in them so students can shake and listen as they try to conclude what is hidden inside. Our local recycling center had a large supply of small cardboard boxes that had once

contained hand soap—voila! We found an ample supply of free mystery boxes! Mystery boxes work well as beginning-of-the-year activities. Using the box, students should first make observations—things they hear or feel. Then they can make an inference—based on the observations, what do you think the object is? Is there any way to know for sure without opening the box? How is this like what a scientist does? This process can help students begin to understand something of the nature of science and what it means to be a scientist.

HOVERCRAFT

Our local recycling center always carries a steady supply of CDs and closable water-bottle tops of different varieties—these materials can be used to make inexpensive hovercrafts. Teachers should build the hovercrafts before presenting them to students for exploration. Using a hot glue gun, teachers glue the base of a water-bottle top that has been cleaned with rubbing alcohol to the center of an old CD (we use the type of spout that pulls up to open and pushes down to close because balloons fit easily over these spouts). When the glue is dry, it is ready for use.

Figure 2

A hovercraft

To operate the hovercraft, students place an inflated balloon over the closed water-bottle top. When the student pulls up on the bottle top, air from the balloon begins rushing out, causing the craft to move.

We've used these models to introduce such concepts as Newton's laws of motion, friction, and force. For example, before the top is pulled up (and opened), we have the students try pushing their devices across the tables. They note how far each device travels without the air rushing through the top and under the CD. We then have the students pull open the top and try the same process. They quickly observe how much farther the device travels when a force—in this case a push—is applied. We then give the students the option to add washers or other weights to see what happens to the distance the hovercraft travels when the same

Connecting to the Standards

This article relates to the following *National Science Education Standards* (NRC 1996).

Teaching Standards

Standard A:
Teachers of science plan an inquiry-based science program for their students.

Standard B:
Teachers of science guide and facilitate learning.

Standard D:
Teachers of science design and manage learning environments that provide students with the time, space, and resources needed for learning science.

amount of force (once again a push) is applied. This exploration leads easily to discussion about Newton's laws of motion.

To extend learning beyond exploration with the simple hovercraft, we often challenge students to find ways to make the hovercraft travel without the students pushing it, or we challenge the students to design a hovercraft that will travel farthest when set in front of a fan in the hallway.

These are just a few of the recyclable items we have adapted for use in our classrooms. We encourage our fellow teachers to visit their local recycling centers to see what types of reusable science teaching treasures they may find. After all, the only thing better than an effective science teaching tool is a FREE science teaching tool!

Reference

National Research Council (NRC). 1996. *National science education standards*. Washington, DC: National Academies Press.

This article first appeared in the Summer 2008 issue of Science and Children.

Chapter 7

Frugal Equipment Substitutions

A Quick Guide

by Erin Peters

In 15 years, I have had science teaching experiences around the country in rural, suburban, and urban schools. In my travels, I have come across some amazingly clever, economical substitutions for hands-on activities in my physical science classes.

Before any substitutions are made, be sure to check with your district science safety officer that they comply with safety standards for your school. The materials can

be purchased at most all-purpose stores. Be sure to make your purchases with funds from the science department budget following all reimbursement procedures for your school. Before soliciting any free materials from businesses in your area, it might be a good idea to get the approval of the science supervisor or principal. I hope that the list below will help those on a shoestring budget provide more hands-on activities for their students.

If You Don't Have . . . Try This!

General classroom materials	
Flexible tape measure	Mark string or yarn • Use one color permanent marker to mark every meter and another color to indicate each centimeter.
Miniature whiteboards	Use Formica sheets • The local hardware store usually sells large sheets of Formica or whiteboard material. • Have them cut the boards into the required size. • Many stores will cut for free if you tell them it is for an educational purpose.
Classroom demonstration balance	Use a large paper clip, tape, and a meterstick • Open the paper clip to look like an *s* and tape the bottom half of the *s* on the 50 cm mark of the meterstick. • If the meterstick is not balanced, attach small bits of clay underneath each side of the meterstick to balance the weight.

If You Don't Have . . . Try This!

Ink for fingerprinting	Use a pencil and tape • Take a fingerprint from a pencil rubbing and preserve it on tape.
Large numbers of "throwaway" items	Find a willing business to contribute • If you need corks, go to your local wine store. • Put your name, contact information, and date of pickup on a bag that is large enough for the items you need, and ask the business owner to put the items in your bag instead of throwing them away. • Pick the bag up promptly and make sure none of the items are contaminated or soiled.

Physics equipment

Spring scale to measure force	Use two paper clips and a rubber band • Slip the rubber band into the two paper clips so you can pull the rubber band by holding the clips only. • Measure the stretch of the rubber band in centimeters. • If you want to find out how much the rubber band stretches for each Newton, hook one paper clip to a ring stand and hook a 1 kg mass to the bottom paper clip; measure in centimeters and subtract the resting length of the rubber band. • Remember that different rubber bands have different amounts of stretch.
Inclined planes	Use metersticks and tape • Tape metersticks together side by side until you have a ramp of the correct width. • Use textbooks to prop ramps up.
Tracks for small cars	Use garden edging • The black edging that is about 10 cm deep works well.
Pulley	Use a dowel rod or pencil • Wrap the string around the dowel rod as you would the pulley.
Electrical wire	Use tin foil • Cut the tin foil into thin strips to act as wire. • Insulate the "wire" with scotch tape.

Chemistry equipment

Phenolphthalein	Use Ex-Lax • Crush up a tablet and dissolve it in 250 ml of isopropyl alcohol.
Spot plate	Make a transparency • Type 12 Os in 72-point font on a sheet of paper. • Spread them out in a 3 x 4 matrix. • Copy them onto a transparency film. • If you have black table tops, you can use a white sheet of paper behind them for maximum visibility.
Pipettes	Use cut-up straws or coffee stirrers • Place the small straw or coffee stirrer into the liquid to be drawn up. • Stop up the open end with your finger. • Release your finger to release the liquid.

This article first appeared in the February 2007 issue of Science Scope.

PART

3

Teaching Strategies That Maximize the Science Budget

Chapter 8

Creative Projects Stimulate Classroom Learning

by Staci Wilson

In a perfect world, what would a good classroom strategy look like? It would have to work for any teacher at any grade level in any discipline; be backed by current research in learning theory; be elegant and simple to facilitate but differentiate for each student; be cost-effective but use a diverse selection of materials; and be active and promote inquiry. Students would be responsible for their own path to learning. They would teach and assess one another, and they would produce original, creative products. Incorporating student projects in your curriculum is one way of achieving these goals.

Classroom projects are original pieces of student work that may be in the form of art, writing, models, experiments, and various other creative outlets. Projects created and presented by students, based on units being studied, stimulate learning and give students the opportunity to follow their own interests. Brain-based learning theories uphold that classrooms that are noisy, active environments where students are engaged in individual learning paths can be conducive to students learning at high levels (Jensen 1995).

With projects, students choose their own path to learning by creating original products that are shared and displayed for others to learn from them. Students take ownership in their learning by creating products that are designed with their own individual interests, talents, and learning styles in mind. Brain-based research also suggests that when students teach what they have learned, they use their whole brain to do it, which makes long-term storage and retrieval of information more efficient.

TYPES OF PROJECTS

Projects come in a series of categories that are inclusive and flexible and offer the opportunity to be creative. Specific examples of projects include solar system models and atom models of all types, such as data collecting for a sunspot activity and posters about the rock cycle or periodic table groups. Artistic designs have included "The Lightning Song" and "The Double Binary Star Dance." Computer applications have included slide shows about native rocks and minerals and the life of Albert Einstein. Students enjoy making games, such as

Further Reading

- "Quest Guidebooks," from the April/May 2007 issue of *The Science Teacher*

PreK-5

Figure 1

Project instructions for students

1. Pick a topic from the various units of study for the upcoming nine weeks, including chapters ___ of the text. These topics include _____.
2. Choose a project category from the following: models, experiments, creative writing, posters, educational tools, classic research, book or article critiques, games, computer applications, or other artistic outlets. Consider your own interests and learning styles. Pick a project that you can enjoy doing, and get approval from the teacher for your idea. All projects must be safe.
3. Sign up for your individual choice on the sign-up sheet located _____.
4. List materials needed. Determine whether the materials are available in the classroom; if not, consider whether access is available outside of school.
5. Bring all materials to class starting on _____. We will work on the projects in class for two blocks (or three 50-minute periods) only; otherwise it will be homework.
6. All projects end with a product, an oral presentation, and a three- to five-minute assessment for the audience.
7. Projects will be due according to when the individual units are close to finishing. All projects on topic _____ are due _____ and those on topic _____ are due _____.
8. The product can be almost anything. Although the challenge level of the products will be scored, the idea is to do something that will create a lasting

memory for what has been learned. Expect to spend four to six hours completing most projects, starting with research, gathering materials, product creation, and preparation of assessment. Do not forget to ask for copies or a transparency if needed.

9. Remember, the product should be a creative way to teach the audience the details of the topic. The assessment should check the audience's learning. The total time of the presentation should not be shorter than 9 minutes or exceed 15 minutes.
10. Example projects for a meteorology unit:
    Models—atmosphere in a box, terrariums
    Experiments—build a rain gauge and measure precipitation
    Creative writing—poem about lightning, short story about a hurricane, children's book about seasons
    Posters—fronts and weather systems, wind patterns
    Educational tools—teach a mini-lesson on pressure using manipulatives
    Classic research—write a report about Ben Franklin or lightning-strike survivors
    Book or article critiques—write a summary about ball lightning and include pictures
    Games—weather trivia complete with game board, instructions for play, and game pieces
    Computer applications—create graphs on global-warming data
    Artistic—write a song about snow, perform a rain dance for a cultural demonstration

"The Great Earth Science Game" and "Chemical Trivia." Students review books like Carl Sagan's *Cosmos* and articles from *Astronomy* magazine. Creative writing comes in many forms, including a short story called "A Trip Through the Milky Way" or a fable about how the universe was created. Educational tools such as study guides and pretests are generally reserved for students considering education as a career. My personal favorite has been the songs. Students have played guitars while they sing and even performed a "doo-wop" song in Earth science that was awesome.

GETTING STARTED

To begin, prepare a list of topics you will be covering in the upcoming term and a list of suggested project

categories, and insert this information into the Project Instructions for Students sheet (Figure 1). Also insert the chapters being studied, sign-up sheet location, the date students should bring materials to class, and due dates. Hand out copies of the Project Instructions for Students to the class. Review the categories to choose from and the topic or range of topics. Emphasize that projects need to be linked to the unit being studied. Students will fill out their topic and category on a designated sign-up sheet. This can be a lined sheet of paper with name, topic, and category sections marked and positioned at a set place in the classroom. You can suggest possible project ideas, or allow students to come up with their own. It is important to limit the number of students who do similar projects; this depends entirely on how much you incorporate the

Figure 2

Rubric for teacher to score each student's oral presentation

Name _____ **Topic** _____ **Category** _____

Product – 35 points

Creative/challenge level	10	_____
Accurate	15	_____
Attractive/high quality	5	_____
Relevant	5	_____
		_____ total

Oral report – 45 points

Voice clear, posture good, not reading	5	_____
Appropriate length of presentation	5	_____
Depth of understanding, mastery of topic	15	_____
Strength of communication, clarity of lesson	15	_____
Can answer questions about the project	5	_____
		_____ total

Assessment – 10 points

Assesses lesson	5	_____
Clarifies concepts	5	_____
		_____ total

projects. Projects can be incorporated one per unit as a culminating event or one per quarter to be presented as the various units are covered.

MATERIALS

Materials can be inexpensive, routine items that many teachers use in the classroom. Items to have in stock include poster board, sketch paper, construction paper, scissors, glue, Styrofoam, cotton, markers, glitter, toothpicks, cardboard, textbooks, magazines, and dice. Of course, the materials can be varied or limited. Students can also be responsible for bringing in their own materials. Your school library can be an invaluable resource for gaining access to magazines, books, and computers. Miscellaneous items can also come in handy, such as packing peanuts or buttons.

PROJECT PRESENTATIONS

Every project ends with an oral presentation. The presenter's goal is to tell the audience what he or she learned or clarified by doing the project. Each presenter can design a three- to five-minute assessment to see whether students in the audience learned or mastered the content from the presentation. It could be a short quiz or game; the range of possibilities is endless. If time is short, the presenter can come up

with three questions for the teacher to use as a unit or topic assignment instead.

ASSESSMENT

Each member of the audience can anonymously score the presenter based on the criteria for a good learning experience, including speaking skills, depth of presentation, and creativity of the product. It can be as simple as ranking the presentations on a scale of 1 to 10, with a comment justifying the score. In addition, you can use the rubric in Figure 2 to rate each presentation yourself.

MANAGEMENT TIPS

When 30 students do 30 different activities, things can get chaotic. The noise level in the room will grow when all the students are engaged. Discuss the importance of using soft voices and limiting movement across the room. Start slowly, with one or two classes, until your organization system is in place. Make sure students understand behavior expectations and the consequences for being off task.

Have students use a sign-up sheet to track the various projects. During sign-ups, have students share aloud their ideas and progress. Although no two students should do exactly the same project, the students can feed off one anothers' ideas as their plans develop. Have a whole-group discussion at the beginning of each class so that questions are answered for everyone. This lowers the demand on the teacher for individual questions and the wait time for students. If you have a larger classroom, consider having students work in teams on projects.

If possible, show projects that have been completed in the past and tell students how they were scored. This can improve the quality of future projects. Also, limit the number of times a student can use the same category per semester. This encourages students to try new categories, which will broaden their horizons.

Make sure students know where materials are located. When most students are close to finishing, ask them to finish up the projects as homework. Allow students to take home any necessary materials needed to complete the projects at home.

During oral presentations, ask the audience to assess each presenter to keep the audience engaged and motivate the presenters. The audience can use the teacher's scoring rubric (Figure 2, p. 43) or simply rate each presentation with a number or letter grade and provide a brief explanation to justify the score.

CONCLUSION

Projects are inexpensive, active, and creative. They shift the responsibility of teaching and learning to the student. Students design their own path to learning, using their individual talents to create and share the way they learn best. The students learn at levels high enough to teach the topic, and they enjoy it. As a bonus, the products become great visual aids for future classes, if the teacher chooses to keep them.

Reference

Jensen, E. 1995. *Superteaching.* San Diego: Brain Store Inc.

This article first appeared in the October 2004 issue of Science Scope.

Chapter 9

A Geometric Scavenger Hunt

by Julie Smart and Jeff Marshall

C hildren possess a genuine curiosity for exploring the natural world around them. Despite new playground equipment and a renovated kickball field, my third graders still gravitate to an outdoor area teeming with vines and other plant life. Warnings to watch out for spiders or other unexpected creatures seem only to further pique their interest to discover the wilds.

One afternoon as I watched my students exploring during recess, I began wondering how I could get them to study mathematics or language arts with a similar enthusiasm. As we lined up, one of my students hurried to bring me a sweet gum ball she had found in the woods, exclaiming, "Look, Mrs. Smart! It's a sphere just like the one we learned about in math!" As the other children crowded around to see her find, one of my students asked a question that would lead us into a four-lesson inquiry investigation that integrated

mathematics and science. He asked, "Mrs. Smart, do you think there are more shapes in the woods?" We were about to find out.

ENGAGING STUDENTS

As my students marveled at the sphere from the woods, I began to see a way to spark student engagement in geometry. My students had already completed a unit of study on geometric concepts in math. During our geometry unit, I had provided many application-based learning activities such as exploring geometric features of our school and designing miniature buildings using specific combinations of geometric figures. However, my third graders still seemed to view geometry concepts as isolated and irrelevant to their daily lives. Integrating science and math in this activity provided an opportunity for my students to apply their knowledge of geometry to real-world situations and extend those mathematical concepts to a new context.

To begin the investigation, I posed a question to the class that was similar to the one the student had asked while outside: "Do you think there might be more geometry in the woods?" Hands shot up in the air as students volunteered other examples of possible geometric finds waiting to be discovered in our schoolyard habitat. One student suggested

---

Further Reading

- "Magnet Scavenger Hunt," from *A Head Start on Science* (2007)

---

we might find leaves shaped like triangles, and another guessed there may be sticks to represent line segments.

After brainstorming as a class for 15 minutes, students took a few minutes to record their predictions and reflect on the day's events in their science journals. I was excited at how engaged my students had become in this endeavor, so I planned to dedicate a second lesson to exploring the environment around our school in what my students had already dubbed "A Geometric Scavenger Hunt." I followed district guidelines for outdoor field trips, which involved addressing these considerations: parental permission; number of chaperones; medical issues (student allergies, sunscreen, medications, etc.); appropriate clothing (long pants, long-sleeve shirts, closed-toe shoes, etc.); communications (available cell phone, two-way radio, etc. in case of an emergency); and review of site (to establish it is clear of hazards, including poisonous plants, insects [ticks, mosquitoes, etc.] and man-made hazards [trash, broken glass, etc.]).

As my students packed up to go home that day, I overheard several lively conversations about the pending Geometric Scavenger Hunt. It had taken a while, but my students were finally "hooked" on geometry.

EXPLORING NATURE USING GEOMETRY

For our second lesson, the exploration phase of our investigation, I divided students into teams of three. I explained that students would be using disposable cameras to document their work (one camera was provided per group). In an effort to maximize potential learning and to keep students focused, I assigned each student one of three roles on the team: Photographer (responsible for taking photographs of geometry in nature), Scribe (responsible for documenting the justification for each photo), or Manager (responsible for having a list of geometry terms and for keeping the group on task during the activity). After assigning teams and roles, I explained each job and made sure that students understood their individual responsibilities before proceeding.

Before leaving the classroom, I reiterated a few key safety issues. First, I indicated the areas where students could explore, which included only areas inside the school property where students would be in my visual range at all times. Although I had already checked for hazards, as a precaution I showed photos of several poisonous plants that are native to our area and reiterated that students were not to handle these plants nor were they ever to eat any berries or other plant parts. I also reminded students to

be mindful of sharp or pointed objects like branches. Finally, I told students they were not allowed to go into any overgrown areas, which could be home to snakes or other creatures. The students were ready to begin the investigation.

Students quickly scattered around the schoolyard as they found and photographed examples of geometry in the plant life around them. Students identified and photographed leaves that were symmetrical, branches that formed acute angles, cylindrical logs, and flower petals flecked with miniature line segments. One group of students pointed out an abandoned bee's nest containing hexagonal chambers, and I photographed it; another group observed wild berries exemplifying spheres.

The effectiveness of teamwork was evidenced by the ownership each child took regarding the assigned task. As I circulated among the teams, I saw students making insightful observations using their knowledge of geometry coupled with their skills relating to scientific inquiry. One student observed, "The veins on this leaf come together to form acute angles." I overheard another student comment on the abandoned bee's nest: "The chambers in the nest are all shaped like hexagons, and they're also the same size so that makes them congruent."

Following approximately 30 minutes of exploration, we returned to the classroom. Students washed their hands with soap and water and checked their clothing for ticks, and then began writing reflections of the exploration in their science journals. The use of reflective journaling was an essential component of this investigation. At the end of each lesson, I provided a writing prompt that encouraged students to examine their own thought processes pertaining to the connections between geometry and nature. The prompt for the second lesson read, "What did I learn today that helps me think about geometry in a different way?" One student responded, "I used to think that geometry was just shapes and lines, but now I see that I can find geometry in the world if I look for it. I know how to name the shapes when I see them on paper, and now I can find them when I see them in real life." This student was becoming aware of his ability to identify geometric figures in a new context. Reflective journaling is a method of chronicling this process of metacognitive growth (Hubbs and Brand 2005).

At the end of the second lesson, I gathered the disposable cameras to have the film developed that evening. The next day students would be able to view their team's photographs and provide a rationale for their geometric classification of items found in nature.

EXPLAINING OUR FINDINGS

At the beginning of the third lesson, I distributed each group's photographs and gave the teams an opportunity to match the photos with their notes from the previous day. Teams then began to share their findings with the class and offer geometric justification for their photographs. Students from other teams questioned and in some cases challenged their reasoning for selecting each item. For example, one student disagreed with a team's classification of a photographed log providing an example of a cylinder because it did not have a circular face on both ends; one end was circular, but the other end was conical or pointed. Eventually, the class decided to exclude the object because it lacked the properties of a cylinder that were learned earlier in the geometry unit.

Other similar discussions helped students confront misconceptions regarding geometric shapes and figures seen in the natural world. For example, one student photographed a long stick and had labeled the object a line. During our discussion, a classmate commented that a line "has to extend infinitely in both directions, and the stick has definite endpoints." The original student then decided to reclassify the stick as a line segment. Another student shared a photograph of a flower that was missing several petals. She commented, "I can tell that this flower used to be symmetrical, but several of its petals have fallen off. Now it doesn't have any lines of symmetry." A classmate then observed that if several more petals were strategically removed, the flower would once again be symmetrical. This process of class discussion helped students make clear distinctions between geometric figures and their properties.

At the conclusion of the third lesson, students again wrote a reflection in their science journals about the day's activities, answering the following prompt: "Write about a time during today's class discussion when you did not agree with a group's classification of its photograph. Did you change your mind after hearing the group's explanation?"

LINKING TO THE ANIMAL KINGDOM

As the third lesson drew to a close, students began wondering about animals and their geometric traits. Having recently concluded a science unit on animal species, adaptations, and habitats, my students were curious to explore representations of geometry in the animal world as well. I incorporated an online extension activity as a fourth lesson. Students explored geometric features in the animal world using several kid-friendly animal webquests *(www.bestwebquests.com; www.webquest.org)*. Each student individually explored these links to observe a wide array of animal species.

Once again, students were responsible for recording their observations about geometry in the animal world and justifying their classification of geometric elements. Students observed circular patterns on a cheetah's coat, the cone-shaped barb on a stingray, the symmetry of birds' wings, and the triangular teeth of the great white shark. Perhaps the most interesting observation came from a student who observed that an anaconda unhinges its jaw to form an obtuse angle to consume its prey whole. This student went on to explain that most other animals are only capable of opening their jaws into an acute angle.

Other students also made inferences that much of the geometry found in the animal world is very adaptive and functional for the animal. During our class discussion following the webquests, students pointed out how some geometric patterns on animals help camouflage them in their natural habitats for protection from predators. One student noted the importance of the symmetry of bird and butterfly wings in the animals' ability to fly, and another mentioned the significance of sharp triangle-shape canine teeth for carnivores. An overarching theme of "functionality" emerged as we discussed the examples of geometry discovered in the animal world.

At the conclusion of the fourth lesson, students wrote a final reflection in their science journals about their experiences with the webquests. As a final extension activity, students were given the choice of creating either a PowerPoint presentation or a poster to highlight several of their findings from the four-day investigation. This open-ended activity provided an opportunity to differentiate by both ability and interest, which has been shown to increase student motivation in the classroom (Tomlinson and McTighe 2006).

ASSESSMENT

In an attempt to authentically assess student work for these activities, I developed a performance rubric to accommodate the various formative learning aspects of the investigation. Each component was evaluated on a scale from 1 (incomplete or missing) to 5 (exemplary). The components evaluated included performance in the assigned role during the outdoor investigation and subsequent team presentation, thoroughness of science journal reflection responses (both science and mathematics concepts), and the correct identification of multiple geometric figures associated with the second lesson and the webquests portions of the investigation.

The student's overall score was then converted into a percentage of the total points available.

Overall, this investigation allowed students to view nature through a different lens, a geometric one. In the days that followed our investigation, students continued to make insightful observations about the geometric figures they saw all around them. Geometry was no longer just an isolated concept in their math books; rather, it provided a tool that allowed them to examine their world in a completely different way.

References

Hubbs, D. L., and C. F. Brand. 2005. The paper mirror: Understanding reflective journaling. *Journal of Experimental Education* 28 (1): 60–71.

National Research Council (NRC). 1996. *National science education standards*. Washington, DC: National Academies Press.

Tomlinson, C. A., and J. McTighe. 2006. *Integrating differentiated instruction and understanding by design*. Alexandria, VA: Association for Supervision and Curriculum Development.

This article first appeared in the October 2007 issue of Science and Children.

Connecting to the Standards

This article relates to the following *National Science Education Standards* (NRC 1996):

Science Education Program Standards
Standard C:
The science program should be coordinated with the mathematics program to enhance student use and understanding of mathematics in the study of science and to improve student understanding of mathematics.

chapter 9

Chapter 10
Making Connections Fun

by Arlene Marturano

Games are a great way to help students make meaningful connections between abstract science concepts and vocabulary. The following are three games I use to help students connect, review, and reinforce what they learn in the classroom.

The game Secrets is similar to the magnetic poetry game that asks you to arrange random word magnets on your fridge to form creative sentences. In my version, the words are carefully selected to focus on a specific area of study and are written on index cards. The object of the game is for students to connect the various words in the correct order to reveal something factual (a secret) about what is being studied. For example, when we study butterflies, I provide index cards with words such as *proboscis, compound eye, nectar, wing, veins, scales*, and *barbs*. I also provide cards containing prepositions, conjunctions, and articles. It is up to students to jot down an appropriate verb on a blank index card and then arrange the cards to create a butterfly fact. For example, students might arrange the cards to say one of the following: "The butterfly has scales on the wings," "The proboscis sips nectar," or "Pairs of wings overlap."

To make a game out of it, divide the class into two teams and have a representative from each group stand at the bulletin board. Flip a coin to see who goes first, and then have that person try to make a sentence using the collection of index cards, which should be pinned to the board. The other team can then challenge the sentence if they think that it is factually incorrect. The other student can then pin up a sentence and have it judged by the opposing team. Give each team a point for a correct sentence or for correctly challenging the other team's sentence. Continue calling up pairs of students until all the cards are used up. Before removing the cards from the board, have students copy them into their notebooks to reinforce the concepts displayed.

Connections is a game requiring students to formulate relationships between pairs of concepts. During a unit on solar energy, for example, students observe the temperature and texture of apple slices baked in a foil-lined funnel cooker and an unlined funnel cooker. I write the concepts *reflect* and *absorb* on index cards, tape them to a marker board, and draw a line between them. The challenge for students is to create a sentence that explains the connection between the two concepts as it relates to the solar-cooking activity. For example, "The foil reflects heat onto the apples, which absorb it and cook." Once everyone completes their sentences, students read them aloud and we discuss if they are correct. Those students whose sentences are approved score a point and move on to the next round. At this point I add another concept to the board, such as *solar energy* or *radiation*, and those students still in the game create a new sentence that connects all three concepts. Those who are out of contention can still participate by challenging the remaining players' sentences. We continue in this fashion until only one student remains or all the concepts have been covered. For review and reinforcement, students should record all of the approved statements in their notebooks.

Pairs of Opposites is a great game to use at the end of a unit to gauge student comprehension or review before a test. To begin, have students divide up into pairs. Give each student a set of prepared index cards with vocabulary related to the unit being studied. The terms must have an opposite associated with

them. (These can be created by the teacher or students, if time permits.) For example, for a meteorology unit opposite terms can include *evaporation/condensation, warm front/cold front,* and *atmosphere/vacuum.* For a unit on matter, you can use *melting point/freezing point, solid/liquid, organic/inorganic,* and *protons/electrons.* Students take turns revealing a term to their partners and challenging each other to come up with the appropriate opposite. Again, points can be awarded for every correct match or challenge, and the approved opposites should be recorded in students' notebooks.

This aricle first appeared in the May 2004 issue of Science Scope.

Chapter 11

Take the Eco-Challenge

by Gregory R. MacKinnon and Colin MacKenzie

Earth is in danger of becoming a barren planet . . . Can your students save the environment? No, it's not a new video game or a doomsday scenario. It's a classroom board game we created to "quiz" our fourth-grade students on environmental science concepts.

"Earth Mission: Rescue" focuses on problems such as pollution and wasting resources—but with an emphasis on the problems' societal implications *and* solutions. Working in teams, students must show a working knowledge of environmental issues and demonstrate environmental awareness so that they can eliminate various environmental problems.

The game is patterned after a popular series of cooperative games created by Jim Deacove for Family Pastimes (*www.familypastimes.com*). To advance, student teams rely on their knowledge of basic science concepts, such as energy or the water cycle. "Winning" means students have collectively saved the environment from destruction.

We've successfully incorporated the game into an environmental science unit. Before playing, students are taught introductory notions of conservation of resources, waste management, and recycling initiatives. After playing the game with students, I have seen their motivation, along with their awareness of important environmental issues, increase.

Figure 1

The game board

We typically use the game at the end of a unit as an assessment tool. Teachers can design questions that review the concepts covered in the unit. The game's

Earth Mission: Rescue
Science and Children edition*

Objective	Game Components
To save Earth from destruction by correctly answering environmental science questions to earn enough "enviro-bucks" to gain "healthy" Earth puzzle pieces and complete the picture in the center of the game board	• One game board (see Figure 1) • Game pieces (e.g., plastic animals or other small objects) • Enviro-bucks (Figure 2) • Question cards (Figure 3) • One die

Game Board Setup

On 56 cm × 71 cm (22 in. × 28 in.) bristol board (available at art supply stores) or cardboard, draw a 10-space-by-10-space grid around the edges. Record in these spaces random environmentally related situations that lead either to a loss of turn or advancement (e.g., "You left a chip bag on the playground. GO BACK 2 SPACES") or a gain (e.g., "You brought a refillable plastic drink jug for your lunch! ROLL AGAIN"). The labeling of the spaces is largely up to the teacher but should include positive and negative outcomes, such as roll again and lose a turn. In addition, some spaces should direct students to pick up "Environmental Question" cards.

In the center of the board, draw (or paste) a gray and brown picture of Earth, representing deforestation, pollution, etc. After attaching this picture, draw a grid of 24 squares (6 rows × 4 columns), each about 6 cm × 4 cm, over the Earth picture. Draw (or cut from a magazine) a picture of a lush, green Earth roughly the same size as the grid. Paste the picture onto construction paper, and then cut it into puzzle pieces to fit over the grid squares in the game board center. These are the 24 puzzle pieces won or lost during the game.

Game Play

Students work in teams, but there is no one "winning" team. The class works cooperatively to complete the Earth puzzle.

To start the game, have each team choose a game piece to represent the team. Then cover half the center grid with pieces of the healthy green Earth puzzle and begin play.

By rolling a die, students move their pieces along the spaces of the board and follow the instructions on the space on which they land.

When students land on an environmental question space, they select a question card from the stack and attempt to answer it. A correct answer gains an enviro-buck. An incorrect answer means the loss of a "green" puzzle piece or no gain. When a team earns three enviro-bucks, they can buy a "green" piece of the planet and place it on the board.

Ending the Game

The game is over when there are no puzzle pieces left on the board and either the ugly, brown Earth is showing (the teams have been unsuccessful) or the lush, green puzzle Earth is complete—the students' cooperative effort has saved Earth!

Variations

• Rather than working in teams, have students play as individuals.
• Apply an additional constraint to the game, such as setting a time limit for answering individual questions or playing the game for a set amount of time and seeing how much of the planet students can save in that time period.
• Have students create the environmental question cards themselves. In addition, during play, student teams may opt to help one another answer questions in return for help on future questions.
• Extend the game beyond the traditional board. Instead of covering the gray and brown Earth in the center of the game board, place a larger picture on the wall in the classroom. Each day, ask one student to share what he or she did for the environment that day or to answer an environmental question. Add pieces of the picture until the grid has been filled by the class.
• Alter the pictures in the center of the game board to reflect a different environmental situation, such as a polluted lake next to a factory with a clean lake and animals as the puzzle pieces. One might also design a board with a rain forest scene or in a body of water. By changing the grid and puzzle pieces, a new setting can be created each time.

*Comfortable for up to 18 players (six groups of three) grades 3 and up (depending on the teacher's questions)

Figure 2

Sample enviro-buck

Figure 3

Question cards

To make the cards:

On multicolor 3 cm x 6 cm cards, write environmental questions on various topics in four categories: Environment, Science, Technology, and Society. Use your science curriculum as a guide to writing questions that reflect the content being studied and the level of your students. There are several Content Standards in the *National Science Education Standards* (NRC 1996) with obvious environmental science connections. You may also find other connections to other areas of your science curriculum. Below are some sample question cards we used in the game with our fourth-grade students.

Sample questions:

Environment

How does smoke from factories affect our lives?

Technology

What are four materials that may now be recycled?

Science

Name three ways energy is obtained from the environment.

Society

What kind of waste is produced by fast-food restaurants?

Connecting to the Standards

This article relates to the following *National Science Education Standards* (NRC 1996):

Content Standards
Standard F: Science in Personal and Social Perspectives
- Types of resources (K–4)
- Changes in environments (K–4)
- Science and technology in local challenges (K–4)
- Populations, resources, and environments (5–8)
- Science and technology in society (5–8)

cooperative approach promotes communication among students. The noncompetitive atmosphere complements the idea that addressing environmental concerns is a societal issue—something we all can work on together.

Try the game on the previous page with your students. Use our questions or modify the questions to suit your specific curriculum. Either way, you'll likely find playing a game is a winning activity for all involved!

Reference

National Research Council (NRC). 1996. *National science education standards.* Washington, DC: National Academies Press.

This article first appeared in the Summer 2005 issue of Science and Children.

Chapter 12

Ecosystem Jenga

by Natalie Umphlett, Tierney Brosius, Ramesh Laungani, Joe Rousseau, and Diandra L. Leslie-Pelecky

Students are often taught that ecosystems are "delicately balanced." But what exactly does this mean? How do we help students relate what they learn in the classroom about ecosystems to the world immediately around them?

As scientists who work closely with middle school students as part of a National Science Foundation–funded Graduate Fellows in K–12 Education program called Project Fulcrum, we have learned that abstract concepts, such as "delicately balanced ecosystem," are often not truly understood. We addressed this concern in a seventh-grade science classroom in Lincoln, Nebraska, by introducing students to locally threatened saline wetlands and the endangered Salt Creek tiger beetle (see Figure 1, p. 56). To give students a tangible model of an ecosystem and have them experience what could happen if a component of that ecosystem were removed, we developed a hands-on, inquiry-

based activity that visually demonstrates the concept of a delicately balanced ecosystem through a modification of the popular game Jenga. This activity can be adjusted to fit classrooms in other regions by focusing on a locally endangered plant or animal, which can be determined by contacting local governmental agencies (e.g., Department of Natural Resources).

PREPARATION

Jenga is a popular block-balancing game sold by most large retailers. Small wooden pieces are stacked together to form a tower. Players remove pieces until the tower falls. This activity works best with small groups (four to eight students), and therefore multiple sets are necessary. The price of a Jenga set is approximately $16–20; however, homemade blocks would also work in this activity. For a class of 24 students working in groups of four, six Jenga sets are needed. Alternatively, if students are working in groups of eight, then three sets are needed. Prior to the activity, we modified the Jenga sets to represent our local saline wetland ecosystem. We did so by painting the ends of equal numbers of blocks red, blue, green, and yellow to represent different components of the ecosystem (both ends of each block should be the same color). Paint pens are an easily

Further Reading

• "Survivor Science," from the May 2004 issue of *Science Scope*

available resource for setting up this activity, and the paint dries quickly. The blocks should be arranged randomly when building the initial tower. The activity worksheet that accompanies this article should be copied and distributed, one per group.

IN-CLASS DISCUSSION (BEFORE ACTIVITY)

This activity can be used as an introduction or supplement to an ecology or environmental science unit. The activity can be used to introduce ecosystem concepts (e.g., an ecosystem has living and nonliving components; there are different types of ecosystems such as tundra, desert, etc.) or to reinforce ones students have already learned about. Our activity focused on the local Salt Creek tiger beetle; however, any plant or animal in any ecosystem may be used.

To begin the activity, have students answer the following questions in a class-discussion format:

1. What are the important resources that all living things need?
2. What would happen if a part of the ecosystem (living or nonliving) was removed permanently from the ecosystem?
3. If a plant becomes extinct, could this cause some animals to become extinct as well? If so, how?

Next, divide the class into groups of four to eight students. Distribute one Jenga game, one die, one Activity Worksheet, and one pencil to each group. Have each group of students set up the Jenga tower three blocks wide, with each level of blocks being perpendicular to the level below it. That is, if the blocks in the first level are pointing north-south, the blocks in the second level will point east-west. Also, make sure that students distribute the different-colored blocks randomly throughout the tower.

Before starting the activity (but with students situated in their groups), have the class discuss what the tower represents. Students should recognize that the tower represents a healthy ecosystem and that each block represents one component of the ecosystem (for example, red = animal species, green = plant species, blue = water, yellow = air). These color assignments are large general categories of ecosystem components and can be designated before the activity and given to students. If there is an organism of interest to the class or local ecosystem, the corresponding block color can be assigned to that organism. We suggest using the large categories, because this shows students that *all* the components of an ecosystem are critical to its health and does not place a disproportionate importance on

Figure 1

Background

Most people associate environmental destruction with coral reefs and rain forests that are (in most cases) located thousands of miles from their homes. Most Nebraska residents are unaware that one of the rarest ecosystems in the world is right outside their back door. Saline wetlands and salt flats historically dotted the landscape of Lancaster County, located in southeast Nebraska. Currently, areas just outside the city borders of Lincoln are the last known locations of the Salt Creek tiger beetle. The Salt Creek tiger beetle is a predatory insect that captures other small invertebrates with its mouthparts. This behavior, and the tendency for these beetles to have stripelike markings, gave rise to the tiger beetle's name. The Salt Creek tiger beetle is an indicator species for these rare saline wetlands: The presence of this beetle indicates that the wetlands are in good health.

Salt Creek tiger beetle population numbers have dropped to fewer than 500 individuals, resulting in the beetle being one of the rarest insects in the world. This drop in numbers led local scientists to work toward getting the Salt Creek tiger beetle onto the federal endangered species list, which was accomplished in October 2005. The beetles' disappearance has been linked to habitat loss from city development, pollution, and pesticides.

Your local National Fish and Wildlife Services can help you identify local ecosystems you can use as the centerpiece of this activity. Making the activity specific to your area helps educate your students about local wildlife, as well as local conservation issues.

any single organism. To begin the activity, have students complete the questions on Part I of the Activity Worksheet. Next, explain the directions to students (see Part II of the Activity Worksheet), have them play the game twice, and then have students discuss their results with the whole class. Guide students through the discussion with the following questions:

4. How many blocks did you have to remove to destroy your tower? Was it the same each time?

Note: Groups will have many different answers. There are no right or wrong answers for this question. Writing each group's results on the board will illustrate that all ecosystems are unique, so no two will be destroyed in the same manner. We conducted this activity in a number of classrooms and on average it took the removal of 12 blocks for the tower to fall; however,

Activity Worksheet

Ecosystem Jenga

Part I

How do we know if an ecosystem is healthy? What would a healthy ecosystem look like?

How do we know if an ecosystem is unhealthy? What would an unhealthy ecosystem look like?

What are some ways in which the nonliving parts of the ecosystem can be damaged?

What does the tower of blocks represent?

What does each color of block represent?

What does removing a block represent?

Part II

Materials (one set per group of 4–8 students)

- Jenga game
- Die
- Activity worksheet
- Pencil

Directions

1. Roll the die. If you roll a 1 or 6, you have not damaged your ecosystem, so do not remove a block. If you roll a 2, 3, 4, or 5, use the table below to determine which block you should remove. Do not take blocks from the top!

2. When you remove a block, set that block aside and make a tally on your worksheet. Do not return any blocks to the tower! *This would be a good time to explain to the students that you cannot return blocks to the tower because, for instance, if an animal becomes extinct, it cannot magically appear again in the ecosystem.*

3. Take turns rolling the die and removing blocks until the tower falls.

4. Record the number of blocks taken of each block type by placing a tally mark in the appropriate box. Also record who pulled the last block from the tower before it fell.

Number on die	Ecosystem damage
1	No damage to ecosystem
2	Water pollution = blue
3	Animal extinction = red
4	Plant extinction = green
5	Air pollution = yellow
6	No damage to ecosystem

Color of block	Type of damage	Number of blocks taken first round	Number of blocks taken second round
Blue	Water		
Red	Animals		
Green	Plants		
Yellow	Air		
Totals:			
1. Type of final block removed		1.	1.
2. Student who removed last block		2.	2.

some groups removed as few as 3 blocks, while others removed 20 blocks before the tower collapsed.

5. Who is to blame for the ecosystem destruction?

Note: Fingers will be quick to point to the last person who pulled a block from the tower; however, everyone who took a block from the tower helped make the tower fall. Everyone involved shares the blame for ecosystem destruction.

6. What would happen if the Salt Creek tiger beetle became extinct?

Possible answers:
a. Maybe nothing. The ecosystem would be weaker, but removing this species of insect might not cause the entire ecosystem to be destroyed.
b. The ecosystem could collapse.

7. Could a plant becoming extinct cause the extinction of the Salt Creek tiger beetle?

Possible answers:
a. Yes, the tiger beetle may eat an insect that feeds on that plant.
b. No, tiger beetles do not eat plants.

Note: Because scientists do not know what would happen if the Salt Creek tiger beetle were to become extinct, there are no incorrect answers to the last two questions; however, students must be able to support their answers with data from their games.

ACTIVITY BENEFITS

This activity allows students to learn about ecosystems in a tactile way. Although a field trip to an actual ecosystem is ideal, not all schools can offer students this opportunity. This activity gives students a concrete and tangible model with which to work. We observed that students were excited to play the game, and even those students who normally were disruptive or unengaged participated in the activity. Aside from our own

observations, a number of teachers also commented that this was the case.

By completing this activity, students

- observed how each trial was different and that it took the removal of more, less, or different blocks to destroy the ecosystem;
- demonstrated how species becoming extinct may weaken, yet not destroy, an ecosystem;
- discovered that the destruction of an ecosystem can't always be blamed on the extinction of a single species;
- became more aware of the local saline wetland ecosystem and the endangered Salt Creek tiger beetle; and
- showed a continued interest in the local ecosystem as evidenced by the articles from the local newspaper that they brought in to share with the class.

These results led students to the realizations that all ecosystems are different and that different ecosystems respond differently to destruction. Although we focused on a seventh-grade class, other teachers at different levels, including first, second, and fifth grades, have adapted the activity for their classes.

EXTENSIONS

Through this activity, students can also learn the value of restoring ecosystems. To model restoration, have students return a yellow or blue block (air and water) to the tower whenever they roll a six. Students see that restoration helps prolong ecosystem life, but if the damage continues, the ecosystem still collapses. You can also incorporate a lesson on the importance of keystone species to ecosystems by starting with a tower that has a combination of red and green blocks at its base. When students roll a one, they must remove a block from the bottom layer, which represents the keystone species. Students discover that the extinction of a keystone species has a larger and quicker impact on the ecosystem than do other species.

This article first appeared in the September 2009 issue of Science Scope.

Chapter 13

Cartooning Your Way to Student Motivation

by Derek Sallis, Audrey C. Rule, and Ethan Jennings

Unmotivated, underachieving students pose a huge challenge for teachers. One way to motivate and stimulate student interest in a topic is to use humor. Humor can help students make new connections in learning and improves retention of information (Garner 2006). In this article, we describe how we integrated art and literature with science to encourage curiosity through the exploration of rocks, crystals, and fossils; to fuel interest with science trade books; and to translate newly acquired science information into funny cartoons.

Analyzing, improving, and creating cartoons related to science content has been shown to increase both science achievement and student motivation in a study of sixth graders learning about rocks and minerals (Rule and Auge 2005). Interestingly, a study on the efficacy of cigarette warning labels (Duffy 1999) showed that the addition of a cartoon figure to the warning increased the believability of health consequences for adolescents. In addition, creativity gives Americans a competitive edge in the global economy against other countries with equally technically skilled workers (Zhao 2008). Yager (2000, p. 337), in his vision of reformed and effective science education for the year 2025, lamented, "Much research and development has been done on developing students' abilities in this creative domain, but little of this has been purposely incorporated into science programs."

The cartoon-making activities described here (see Activity Instruction Sheet, p. 63) engage students while they learn Earth science concepts and develop their abilities to visualize and combine ideas in new ways. Our diverse classroom population of African American, Latino, and white middle school students enrolled in a class for reading enrichment and improvement enjoyed the activity, remaining almost glued to the tasks for the entire class period. (See the cartoons they created, Figures 1 and 2, pp. 60 and 61.)

ENGAGING STUDENTS THROUGH SIGHT AND TOUCH

We began by providing hands-on materials to focus student attention and to engage tactile learners. We provided rock samples with interesting textures, glittering

Further Reading

- "Cartoons—An Alternative Learning Assessment," from the January 2008 issue of *Science Scope*

Figure 1

Polar pinball

POLAR PINBALL
Try Not to Melt The Icecaps

I hope I don't lose and melt!

Global climate change brings dangerous changes to Earth. The polar ice is already melting.

crystals, and "cool" fossils such as a reproduction of a dinosaur footprint and some intricately detailed horn corals. We also provided vesicle-filled lava rock samples and a stalactite from the blast zone of a remodeled commercial cave entrance (remind students to never touch or remove stalactites from natural caves). Eye-catching mineral samples and fossils are often sold in museum gift shops, science stores, and rock shops. Alternatively, you could invite a guest speaker to visit and show specimens from a collection.

Allow students, who are seated in small groups for all of the activities described in this article, a few minutes to examine the materials and talk to peers about what they notice or know about them. Ask students to sketch two specimens that they find interesting and record their observations. Call on volunteers to share observations and infer the origin of the specimens, providing additional information as needed so that all students have a basic understanding of the main features of the items.

Our students were fascinated with the materials. They recognized the dinosaur footprint but were less able to identify the other fossils. This initial formative assessment provided us with information about students' background knowledge so that we could better support their learning. Because our work was conducted as guests in a remedial *literacy* (rather than science) class, we did not have the opportunity to conduct earlier lessons to build a stronger foundation for learning about these Earth science topics; however, many of our students made connections to previous science lessons. You may want to extend this lesson over several days with your students.

READING EXCITING TEXTS

After student interest was piqued through examination of specimens, we provided Earth science content information to them through richly illustrated texts written at a range of grade levels. We used nonfiction picture books written for elementary and middle school audiences, although more technical texts, field guides, or coffee-table books can be used if they have colorful, exciting pictures. Because this lesson took place in a class for struggling and easily frustrated readers, it was important to provide highly visual books with short text passages so that comprehension was supported with images.

We brought a large selection of books so that students would have ample choices from which to find a book that ignited their curiosity and presented textual information at an independent reading level (hence the large number of elementary books). We used books that focused on minerals and crystals, volcanoes, earthquakes, glaciers, caves, and dinosaurs or other fossils. In this way, students could choose books that focused on their particular Earth science interests. You might supplement what is available in your school library with books that you check out from your local public library for this in-class activity. Be sure to provide enough books so that students have a lot of choice and can switch to other books when they have finished reading or if they realize that the book is too difficult or not of interest. Ideally, have twice as many books as students.

We asked students to find a book of interest and to read silently for 20 minutes. To help them quickly choose appealing books, we sorted the books by topic and connected these with many of the specimens students had just explored. Our students were eager to browse the books because of their interesting illustrations, some of which matched the samples we had supplied. They began by paging through and reading excerpts; soon they were fully engaged in reading. They read silently and individually for 20 minutes, pausing occasionally to show classmates interesting facts or images in the books. After 20 minutes, we asked students to talk to group members about what they had learned. Then students recorded a few facts, using complete sentences, in their notebooks.

ANALYZING EARTH SCIENCE CARTOONS

After they looked at the books, we told students that soon they would make cartoons that incorporated content related to minerals and crystals, volcanoes, earthquakes, glaciers, caves, or fossils, using humor to convey their newly acquired Earth science content from the reading. To prepare for the cartoon making and to inspire students, we showed them colorful, funny cartoon examples related to these same Earth science topics made by preservice teachers (Rule, Sallis, and Donaldson 2008), available free online (see References for a link to the article and cartoons). Each of these cartoons is accompanied by an explanation of the pertinent science content. We printed a color set of the cartoons with their explanations from the article appendix and circulated them among groups of students.

We gave students time to read the cartoons and enjoy the humor. Then we asked each group to choose two cartoons and explain to the rest of the class the science content and sources of humor. Humor often depends on wordplay such as homophones (e.g., *dear* and *deer*) or words with multiple meanings (e.g., *play*: to toy with an object, a dramatic skit, or a sports move), exaggerated emotions, impossible situations, or parody. Some classic books that illustrate common puns are the series written and illustrated by Fred Gwynne who played Herman Munster in the popular TV series *The Munsters* (see Resources for more information). If time permits, you may want to use these and other books suggested by your language arts colleagues or school librarian to build a foundation for recognizing and creating humorous wordplay.

MAKING ORIGINAL CARTOONS

The goal for this activity is to help students take Earth science concepts that they have learned through reading nonfiction trade books and to re-communicate them to others through the medium of humorous or clever cartoons. The NSES Science Content Standards state that more emphasis should be placed on "communicating science explanations" (NRC 1996, p. 113). This cartoon-making activity allows students to integrate literacy, science, and art for the purpose of communicating science learning. As in the other activities, students should be seated in small groups of three to five. Students may work individually on a cartoon or with one or more members of the group. Each student or team of students might begin a cartoon and continue working on it with input from other group members.

To ensure that a solid content foundation is being established, ask students to begin by writing a statement of the science information that will be conveyed by the cartoon they make. This might be one of the fact statements that they wrote after reading a trade book. Sometimes, as in the "polar pinball" cartoon

Figure 2

Glacial meltdown

Glenda Glacier was not at all pleased with the forecast for more global climate change.

example (Figure 1), this statement becomes the caption of the cartoon. Other times, the science ideas are inferred from the cartoon visuals and caption. We exposed our students to a wide range of Earth science topics through the specimens, trade books, and example cartoons because we wanted them to choose topics of high interest for reading and designing cartoons. You may want to reduce the number of topics to fit more closely with what your students are studying. The example cartoon resource discussed here provides several samples for each of the six topic areas; therefore, students will have models even if your focus is narrower.

As stated in the *National Science Education Standards'* Science Teaching Standards, "Student understanding is actively constructed through individual and social processes" (NRC 1996, p. 29). A student clarifies thoughts while explaining them to others and receives helpful feedback and fresh ideas from classmates. Students may want to brainstorm, in groups or as a whole class, science words related to what they have just read that have homophones or multiple meanings to help in creating humor. For example, some possible terms for wordplay are *order, quartz, crystal, model* (crystals); *Old Faithful, explode, dormant, sleeping, erupt* (volcanoes); *Richter scale, fault, shocking, shaker, crack-up, love waves* (earthquakes); *calving, freezing on, terminal moraine* (glaciers); *cast, mold, bone to pick* (fossils); and *deposit, bank, cave, batty* (caves). Identifying the puns used in the example cartoons from "Humorous Cartoons Made by Preservice Teachers for Teaching Science Concepts to Elementary Students: Process

Figure 3

Background examples

What type of books would your Earth feature check out of the library?
What happens when your Earth feature goes to the library?

If your Earth feature were a rock star, what hit song would he/she have? What would your Earth feature do on stage?

and Product" (Rule, Sallis, and Donaldson 2008) can also be beneficial. We didn't have any problems with inappropriate jokes or drawings, but you may want to remind students of the limits of acceptable humor: No cartoons that insult classmates, teachers, or groups; no foul language; no violence or sexual content.

Provide color printouts of background scenarios on which students can add hand-drawn characters, talking bubbles, details, and captions to make their cartoons, like those shown in Figure 3. We used clip art in PowerPoint to make familiar or fantasy scenes of human experience and then printed them for students: a hospital operating room, a fortune-teller, a birthday party room, a pinball arcade, a snack shop, a UFO with a beam extending to the ground, the stacks in a library, a talk-show host on the stage, an open scrapbook page, the dance floor in front of a rock band, and an open tent at a campground. To save paper, you might provide each group of students with a set of two to four backgrounds, rather than printing them all for individual students. We found that providing prompts for each setting helped students generate ideas. For example, with the scrapbook setting, we asked, "What Earth science event would your Earth feature want to remember with a scrapbook page?" Students then decorated the scrapbook page, perhaps including a definition of a science term for their personified Earth feature. For a talk-show setting, we asked, "What funny things happened or were discussed when your Earth feature appeared on a TV talk show?" Also allow students to generate their own original settings for cartoons. Supply colored pencils to make the task more interesting and the results more colorful; having erasable colored pencils so that students can easily refine their work is even better.

Students should be allowed access to the texts throughout their cartoon work to find additional information, check facts, clarify ideas, or find details. As students work, misunderstandings they have about their newly acquired science knowledge may appear in their cartoon drawings or captions. An article in *Science Scope* (Song et al. 2008) detailed several ways to use cartoons in assessing student science understandings. As recommended by the authors of that article, classmates might explain their alternate understandings of "controversial" cartoon situations (cartoons showing science facts or processes about which there is disagreement among students as to correctness) and offer evidence from texts to support their reasoning.

As you interact with students, tell them specifically what they are doing effectively and make suggestions for additional components they may want to add to make the concepts clearer. The work can be started on one day and completed at a later class period or for

Overview: In this art-literacy-science integrated activity, students explore Earth science objects, read informational books to learn more, examine humorous science cartoons others have made, and then create funny cartoons of their own that communicate science facts.

1. Engaging students through sight and touch
Materials: Several rock, mineral, and fossil specimens with interesting textures, lusters, or features
Objective: Students focus their attention on Earth science while making observations of specimens and discussing their knowledge or interests with peers.
Time: 10–15 minutes

Provide specimens of interesting rocks, minerals, and fossils to students seated in small groups. Students examine items and suggest to classmates what they are and how they formed. After activating this prior knowledge, volunteers tell observations and inferences to the whole group. The teacher or guest speaker fills in additional information about the origin of each specimen and answers questions.

2. Reading exciting texts
Materials: A large variety of richly illustrated nonfiction books at elementary and middle school reading levels that focus on Earth science topics
Objective: Students read silently and independently for 20 minutes and then record several sentences of Earth science facts that they learned from reading.
Time: 25 minutes

Students choose books of interest that focus on various Earth science topics, including minerals and crystals, volcanoes, earthquakes, glaciers, caves, and dinosaurs or other fossils. They read silently for 20 minutes. Then they tell students in their groups two or three interesting facts that they learned and record these ideas in their notes.

3. Analyzing Earth science cartoons
Materials: Color printouts of the cartoons and explanations from the appendix of the Rule, Sallis, and Donaldson article (2008)
Objective: Students examine cartoons stating the science content and the source of humor.
Time: 15 minutes

Pass out color copies of the cartoons with their accompanying science content explanations. After students have had a chance to look at the cartoons and read the science explanations, assign two cartoons for each group to analyze for science content and sources of humor. Groups report this information to the class. Humor often depends on puns (homophones, words with multiple meanings), parody, exaggerated expressions, and ridiculous situations.

4. Making original cartoons
Materials: Clip-art background scenes for cartoons, erasable colored pencils
Objective: Students write a science fact to be communicated by a cartoon. They choose a cartoon setting and add details to complete a humorous cartoon.
Time: 15–30 minutes

Ask students to record the fact(s) that they would like to illustrate their cartoons. Provide cartoon setting ideas with clip-art scenes and accompanying questions. Students should tell group members their ideas and obtain additional suggestions of puns or humorous situations from them. Remind students of the limits of acceptable humor: No cartoons that insult classmates, teachers, or groups; no foul language; no violence or sexual content. Then students should choose a setting and add characters, details, call-outs, a caption, and a title.

homework, as students often need incubation time for generating creative ideas.

SHARING AND ASSESSMENT

It's nice to share completed cartoons, perhaps by posting them on a bulletin board, so that students can demonstrate their accomplishments. Because of limited access to computers, our students turned in hand-drawn work and we translated their sketches to clip art or scanned and traced their images to make the final cartoons shown here. However, there are electronic ways for students to produce cartoons that can be effective if they have access to software: Students might add to given clip-art backgrounds in Power-Point by using drawing tools or additional clip art, or use other drawing applications, such as Comic Life, to create original cartoons.

A rubric for scoring student work might include the following criteria: (1) statement of science content displayed by the cartoon; (2) explanation of why the cartoon is funny or clever, perhaps including a play on words; (3) sufficient details drawn (talking bubbles, characters, captions, color, elaborations) to make the cartoon understandable; and (4) visual appeal.

CONCLUSION

This technique was successful in motivating under-achieving students to read science books and practice communicating the information by drawing humorous cartoons. The teacher remarked that students were exceptionally on task during the entire lesson. Students put forward consistent effort during the entire lesson—even students who couldn't sketch gave input to partners who did the drawings or wrote descriptions of what was happening in the cartoons. Students enjoyed using their artistic abilities to show their newly acquired knowledge of Earth science topics.

References

Duffy, S. A. 1999. Cartoon characters as tobacco warning labels. PhD diss., University of Illinois at Chicago.

Garner, R. L. 2006. Humor in pedagogy: How ha-ha can lead to aha! *College Teaching* 54 (1): 177–180.

National Research Council (NRC). 1996. *National science education standards*. Washington, DC: National Academies Press.

Rule, A. C., and J. Auge. 2005. Using humorous cartoons to teach mineral and rock concepts in sixth grade science class. *Journal of Geoscience Education* 53 (5): 575–585.

Rule, A. C., D. A. Sallis, and J. A. Donaldson. 2008. Humorous cartoons made by preservice teachers for teaching science concepts to elementary students: Process and product. Paper presented at the Annual Graduate Student Research Symposium, Cedar Falls, IA. (ERIC Document Reproduction ED no. 501244)

Song, Y., M. Heo, L. Krumenaker, and D. Tippins. 2008. Cartoons: An alternative learning assessment. *Science Scope* 31 (5): 16–21.

Yager, R. E. 2000. A vision for what science education should be like for the first 25 years of a new millennium. *School Science and Mathematics* 100 (6): 327–341.

Zhao, Y. 2008. What knowledge has the most worth? *School Administrator* 65 (2): 20–27.

Resources

Gwynne, F. 1982. *The sixteen hand horse*. New York: Bookthrift Company.

Gwynne, F. 1988. *A little pigeon toad*. New York: Simon and Schuster Books for Young Readers.

Gwynne, F. 2009. *A chocolate moose for dinner*. New York: Paw Prints.

Gwynne, F. 2009. *The king who rained*. New York: Paw Prints.

This article first appeared in the Summer 2009 issue of Science Scope.

Chapter 14

Science Newsletters

by Melissa Nail

Having students write and publish their own newsletters is a great way to integrate reading and writing, infuse technology, and build home-school relationships. These newsletters can be used to keep parents informed of what is being taught in class, important test dates, homework and project due dates, and any other information you'd like to share. Involving students in the creation of the newsletters increases their feeling of ownership and reduces the chances that the newsletter will end up in a wastebasket or forgotten in a locker.

GETTING STARTED

My students write their articles using Microsoft Word and lay out the newsletter using Microsoft Publisher. Publisher, which is included in the Microsoft Office suite of programs, is very helpful because it provides a Newsletter Wizard tool that walks students

Further Reading

- "Extra! Extra! Learn All About It," from the November 2007 issue of *Science Scope*

through the steps of setting up a newsletter. All you really need, however, is some kind of word processing software. Today's software allows you to set up columns, place text, import images, and change the type and size of fonts to create a very attractive newsletter. Your students are probably already familiar with how to set up and use word processing software, but you should allow time for a review of the basics. Before they can create their newsletters, students will need to know how to open, edit, and save documents within the word processing program. They will also need to know how to copy and paste text and place graphics in their documents.

PUTTING IT ALL TOGETHER

Once students are familiar with the software, the focus can turn to the content. To get things started, I ask students to brainstorm topics for articles. If they need some guidance, I suggest a few topics, such as explanations of content studied in class, reports on class activities, or summaries of experiments. Topics also can include reminders about upcoming holidays and school closings; reports on new equipment, animals, or other resources in the classroom; and news about the science club, field trips, or after-school events. I contribute a piece to each newsletter called "Teacher's

FIGURE 1

Sample student newsletter

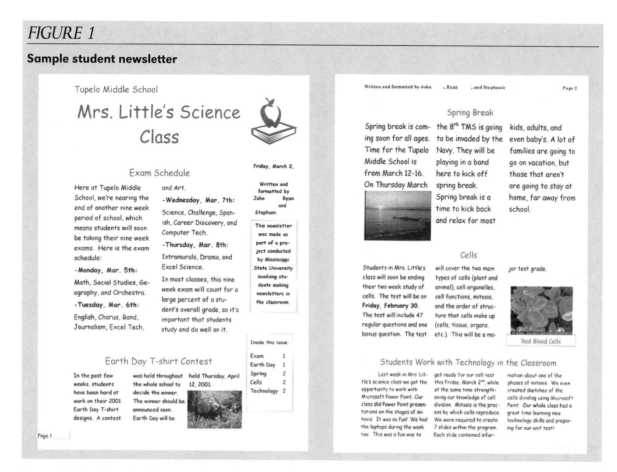

Corner." I distribute it electronically and each group incorporates the column into its layout. I use this column to communicate with parents regarding dates of exams, topics for home study, requests for parent speakers or presenters, and suggestions of ways parents can participate in their child's education.

The internet is the primary source of graphics used to accompany the articles, but students can also create their own figures and illustrations using drawing tools included with the word processing software or other programs. Inexpensive collections of images on disk can also be found at your local computer or electronics store. If a scanner is available, students can also scan in images they have drawn freehand or found in books or magazines. When a particularly good graphic related to a topic is identified or created, such as a student's illustration of a plant cell drawn using Paint Brush, students are encouraged to share it with the other groups.

There are a number of options available to classroom teachers for putting together the student-created newsletters. If wireless laptops are available in the school, the teacher can simply check out the laptop cart for this project. If the school has a computer lab, the classroom teacher can schedule the lab for the creation of the newsletters. The newsletters can also be created in classrooms

with only one or two computers available. This option might require more planning and more time because the students would have to alternate and rotate use of the available computers, with only one group at a time publishing a newsletter while other class members work on other exercises. Some schools now have classroom sets of PDAs that teachers can check out and use. If this is the case, students could use the PDAs to individually input their articles using a program such as Word to Go, Hot Sync the PDA with a desktop computer, and quickly and easily insert the individual articles into their group's newsletter. Some teachers may also wish to collaborate with colleagues within the school so that the students are writing the articles during a language arts lesson and putting together the newsletters during a technology or computer lesson.

In most cases, a cooperative team of students can create a two-page newsletter in only three class sessions. One session is used for brainstorming, planning, and initial prewriting activities. The second session is used for writing and editing articles, and the third session is used for final editing and inserting the articles into a newsletter template. The most time-consuming part of the process seems to be finding appropriate graphics. Selecting one member of each team to be responsible for illustrations or compiling a shared library of appropriate illustrations seems to be a beneficial strategy for speeding up the production process. Another time-saver (especially as students are initially learning to produce their own newsletters) is for the teacher to either create or designate a particular newsletter template to be used.

Each group's newsletter is printed on the school's printer and duplicated on the school copier to make a copy for each member of the group. Unfortunately, our budget did not allow for each student to get a color copy of the newsletter. However, students can take the original file home and print their own color copies if they have access to the necessary software and hardware. The file can also be e-mailed home for students and parents to view in color onscreen. Many students will have their own personal e-mail addresses you can use. Permission should be obtained in advance, however, before you collect and use any parent's e-mail address. Before making copies of the newsletter for distribution, I review the content of each one. This usually takes me about a day.

FINAL PAGE

The student newsletter project proved to be very successful. Students are actively involved in the publication of the newsletters, displaying more excitement for writing articles than they had ever demonstrated for merely writing reports on topics studied. When the finished newsletters are distributed each quarter, students receive them excitedly. They show off their finished products to classmates, then carefully put the newsletters in folders and book bags to take home and share with their families.

This article first appeared in the Summer 2005 issue of Science Scope.

Chapter 15

The Station Approach

How to Teach With Limited Resources

by Denise Jaques Jones

Several years ago, my middle school experienced a huge growth spurt. Before I knew it, my classroom was bulging with many more students than resources. I was determined not to let my now-stretched resources keep me from my favorite labs and computer activities.

My colleague Sarah Harashe and I were able to do this by designing a strategy we call "The Station Approach." We thought this would achieve our purpose, but what we discovered really surprised us. Students loved the stations and continually asked when they would get to do them again. The method increased students' interest in the topic, kept them motivated, and eliminated many behavior problems.

WHAT IS THE STATION APPROACH?

The Station Approach is a method of instruction in which small groups of students move through a series of learning centers, or stations, allowing teachers with limited resources to differentiate instruction by incorporating students' needs, interests, and learning styles. The Station Approach supports teaching abstract concepts as well as concepts that need a great deal of repetition. Stations can cover a single topic, such as density, or several independent topics, such as review-

ing the scientific instruments. Stations can last one class period or several.

The Station Approach is actually an adaptation of the reading groups used in elementary school classrooms. The difference, however, is that in the elementary school model students rotate only to those stations that meet their specific learning needs, while in our approach every student rotates through each station and performs all the activities. Perhaps the greatest strength of the Station Approach is that it incorporates many concepts used for differentiated instruction.

DESIGNING STATIONS AND SETTING UP THE CLASSROOM

Two to four stations are optimum for most activities. More stations can be designed when introducing or reviewing multiple concepts or if class sizes are large. Stations should be independent of other stations and can be completed in random order. When working with large classes, or when using a small number of stations, consider setting up multiples of the same activities and divide your class into two or more rotating groups. Student groups should consist of no more than four to six members. Larger groups have a tendency to become loud and disruptive to other stations.

Design stations so that only one requires the teacher's continued presence. The remaining stations should be self-explanatory or require only limited instructions, which can be posted at each station.

Strive for activities that last approximately the same amount of time. Choose your main activity and modify the remaining activities to take about the same amount of time to complete. The amount of time for each station can vary anywhere from 20 minutes to an entire class period if the content requires several class periods to complete. Designing stations in this way allows for smooth transitions, reducing student frustration at either leaving work incomplete or having to wait idly for other stations to become free.

Design stations so that they vary based on students' different learning styles, interests, and levels of readiness. Each station should require students to look at the concept in a different way. This can be accomplished by thinking of each station as a specific learning style. One station could be your hands-on or kinesthetic station. Here students could complete labs or build models. Another could be the visual station, where students could quietly read, complete computer research, or explore concepts visually. Another station could be an auditory station, where students could have discussions or listen to information on tape.

When setting up the classroom, make use of all available resources—books, computers, and lab equipment. Look around your classroom; something as simple as a lab counter, rug, bookcase, or teacher's desk can become a station. With the Station Approach, even a single computer can become a viable station.

Divide the classroom into discrete areas or stations (see Figure 1). Stations should be flexible and easily accessible by students and the teacher. Stations are not designed to be permanent. They should be easily set up and then removed. Be creative. A bookcase or small rolling cart can change a space into a workable station. Remember to place computers and equipment so that you can easily observe students working. Consider the traffic pattern when setting up your stations. Visualize how students will move from one station to another. Think of your room as a rectangular clock and have students move clockwise around the room. Check with your building custodian or supervisor regarding fire and other safety codes. It is important that you carefully examine all stations for any safety risks. Never block exits, leave extension cords lying where students can trip, or leave objects such as hot plates where they can be knocked over.

Figure 1

Stations and rotation pattern for Matter in Motion

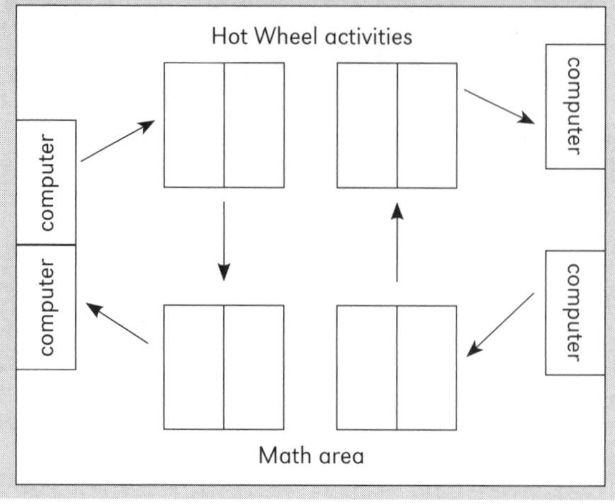

POSSIBLE STATIONS

Lab area—A table or tables enable students to perform hands-on activities.

Quiet work area—Set aside a place in the room (such as a corner) with chairs or a rug. Students sit and quietly complete higher-level thinking questions, peer edit, read supplemental materials, and practice math and process skills. A quiet work area can include televisions or tape players with earphones or books on CD.

Computer—Students can conduct research, design presentations, complete webquests, or use probeware.

Teacher-directed area—Small-group remediation, acceleration, or student-teacher discussions about assignments and projects can take place at the teacher's desk, or by grouping several student desks together in an easily accessible area of the room.

Production area—This may be a section of a lab counter or a table with markers and poster board for easy use by students. If you have AV equipment available, section off a corner of your room and place a video camera, VCR, television, and computer there. Students can record weather reports or perform skits and present them to the class. Solicit parent volunteers or teacher assistance to help students videotape and/or edit their presentations.

MANAGING THE CLASSROOM

When we designed our first stations, we believed they would increase student achievement because we were teaching a concept in multiple ways and addressing students' unique learning styles. In addition, stations allowed us to differentiate instruction by grouping students based on pretests. This allowed us to remediate certain students while accelerating others. What we had not anticipated was the reduction in behavior problems that we observed using the Station Approach. We concluded that this was due to several factors.

First, students weren't required to remain at a task for too long. By rotating students through stations that varied between quiet, mental tasks and active, verbal ones, we kept them interested and reduced off-task behaviors. Students seemed better able to stay quiet and focused when necessary, knowing that they would move soon to a more active station. For some students, the transitions between stations gave them a quick mental break, and just by moving around the room, they exhibited renewed energy and focus.

With this in mind, I have outlined some key points that will help manage the movement around the room and reduce behavior problems.

Student preparation—Prior to using stations, it is important to teach students how the process will work.

Figure 2

Rules during station work

1. Do not disturb the teacher when he or she is working with other groups.

2. Do not leave your station without permission. The only students with permission to leave their group are the supply person and the information person.

3. If you have a question,
   a. Reread the data sheet.
   b. Ask someone in your group.
   c. Quietly ask someone who has already been to that station.
   d. Write your question on a piece of paper and continue to work (if you can't continue, begin a sponge activity, see p. 72) until you see that your teacher is no longer working with another group and is free to help you.

Make sure students understand that all of them will complete each activity no matter where they begin. Be sure to explain how they will move through the series of stations. This will reduce student confusion when it is time to rotate. Let students know ahead of time how long they will spend at each station. Using a large

Figure 3

Posted job assignments, station rules, and station rotations (color-coded). Velcro makes changing jobs and stations easy.

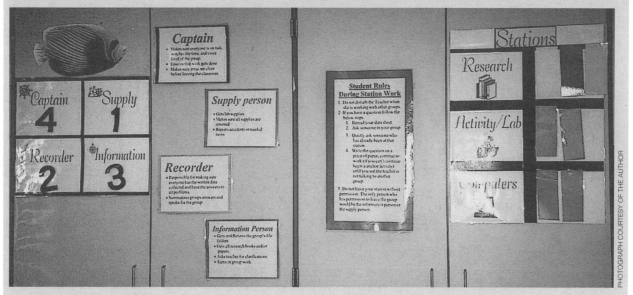

PHOTOGRAPH COURTESY OF THE AUTHOR

Figure 4

Job assignments

Captain

Your responsibilities include

- making sure everyone is on task,
- watching the time and voice level of group members,
- making sure that the group's work is completed, and
- supervising the cleanup of stations prior to rotating to the next station.

Recorder

Your responsibilities include

- completing all worksheets while the group is completing cooperative activities,
- summarizing for the group the decisions or findings that were reached by consensus, and
- speaking for the group when the group is called upon.

Supply person

Your responsibilities include

- getting supplies for the group,
- returning all supplies when finished, and
- reporting accidents or needed items to the teacher.

Information person

Your responsibilities include

- getting and returning the team's file folders,
- getting all research books and/or papers,
- asking the teacher for clarifications, and
- turning in group's and team's individual work.

egg timer will help everyone keep track of time. This helps students adjust their pace and frees you from constantly watching the clock.

Rules and procedures—Display a list of rules and procedures for students to follow. You can post this in the front of the classroom or at each station. Review these rules prior to the activity (Figure 2, p. 71).

Coded stations—Design a rotation strategy that can be displayed in front of the classroom for easy reference. Color-coding or naming areas reduces student confusion (Figure 3, p. 71). Poster board with Velcro numbers and colors makes rotations easy to manage.

Student jobs—When stations have cooperative activities, groups of four work best. Assign each student in the group a job. This will give each student a reason to stay focused and a sense of purpose. Possible jobs include recorder, information person, supply person, and group captain (Figure 4).

Handouts and paperwork—When determining how to handle student paperwork, you may want to consider the following options:

- Place copies of all worksheets and directions at each station. Depending on the type of station, you may require a single worksheet for the group to complete or have each student complete his or her own worksheet. No matter what format you

choose, make extra copies in case of the inevitable lost or messed-up worksheet.

- Copy all necessary worksheets and directions in the form of a packet and hand it out to students at the beginning of the activity. The benefit to this strategy is that it is neat and organized. The only problem is that if a student loses his or her packet, then all of the work is gone, not just the work from a single activity. Designating a crate where students store their packets can minimize the risk of this occurring.

- Create student contracts, which allow the teacher to differentiate instruction. After pretesting, students' work can be modified based on their current level of understanding. A contract is a way of making an individual educational plan for each student based on their pretest scores. Stations can then be written at multiple levels of difficulty, allowing all students to experience success.

Sponge activities—A *sponge* is any activity designed to fill the extra time when students complete their station early or for some reason cannot continue working at their station (see Figure 2). To discourage students from rushing through their work, sponges can include self-checking completed work. This can be accomplished by keeping an answer key in a folder students can obtain after work is completed. Students check their work and correct existing errors. Students must show that work is completed in detail prior to getting the answer key. Another sponge activity is allowing students to peer edit. Independent activities such as Sudoku or logic puzzles can also be made available. Again, students must demonstrate that they have completed the assignment prior to doing these activities.

DESIGNING STATION ACTIVITIES

There are many ways to determine which lessons to turn into stations and how. For the Matter in Motion unit, students stay at each station for 30 to 45 minutes, two sets of three stations are used, and students are grouped (four per group) based on pretest scores on simple speed problems.

Step 1. When designing your stations, first complete a KUD chart. This is a simple unit outline identifying what you want students to know, understand, and be able to do (see Figure 5).

Step 2. Using your KUD chart, determine which concepts you think will give students the most difficulty. From experience, these concepts are usually those that involve mathematical equations and those that require higher-level thinking skills. Star these on your KUD chart. The starred items become the focus of the stations. In the Matter in Motion unit, I knew students would struggle with the speed, momentum, and acceleration problems.

Step 3. Brainstorm all of your potential resources: lab activities, webquests, books, AV equipment and other media, and computers. Make sure to take advantage of resources available through your school library. Think of labs and activities you have done in the past, and those you couldn't do because there wasn't enough equipment for all students. Brainstorming the concept *speed*, I wanted to work speed problems using real-life applications. I determined that planning a trip using different modes of transportation might interest students. We also wanted students to race Hot Wheel cars, despite having a limited number of ramps and stopwatches. These ideas became the foundation for our Matter in Motion stations.

Step 4. Once you determine what is available, decide which activities you want to turn into stations. Look at all of your choices, and pick those that address your topic in multiple ways. Remember that only one station should require a teacher's continued presence. The remaining ones should be designed so that students can complete them independently or with a minimum

Figure 5

KUD chart for Matter in Motion

At the end of this unit students will

Know
- Vocabulary: *motion, reference points, speed, velocity, acceleration, force, net force, balanced force, unbalanced force, gravity, mass, weight, friction,* and *types of friction (sliding, rolling, fluid, and static)*
- Newton's three laws of motion

Understand (generalizations)
- We measure the motion of an object in relation to a point of reference.
- Distance, time, and direction are used to measure motion.
- Unbalanced forces change the motion of an object; balanced forces do not change the motion of an object.
- Friction is a force that opposes motion.
- The gravitational forces of the Earth affect motion.
- The forces of gravity increase as mass increases and decrease as distance increases.

Be able to do
- Identify the relationship between motion and a reference point.
- Identify the two factors on which speed depends.
- Describe the difference between speed and velocity.
- Analyze the relationship between velocity and acceleration.
- Interpret a graph of acceleration.
- Give examples of different kinds of forces.
- Determine the net force on an object.
- Compare balanced and unbalanced forces.
- Explain why friction occurs.
- List the types of friction and give examples of each.
- Explain how friction can be both helpful and harmful.
- Explain the law of universal gravity.
- Describe the difference between mass and weight.
- Carry out an investigation using scientific process skills.

of instruction. You may need to modify these activities so students can complete them in approximately the same amount of time. After determining our foundation activities, we decided to have two sets of three stations. Twelve students in groups of four would rotate through one set of three stations and the remaining 12 students would travel through the second set of stations. The three stations were (1) Hot Wheel car races, (2) computer-aided trip comparisons, and (3) teacher-assisted math problems related to speed.

Step 5. Write station directions, making them as simple and clear as possible. To reduce the amount of paper used, you may want to laminate directions and place them at the stations. This is an opportunity to differentiate instruction by level of difficulty, student interest, or learning style. We decided to use pretest scores as a basis for differentiating our speed problem station. We did this by modifying our speed worksheets so students completed the same number of problems, but at different levels of difficulty.

Step 6. Divide students into groups, determined randomly, using pretests for readiness or by using interest surveys. Group sizes will ultimately depend on the availability of equipment and class size. If you have three computers available, placing students in groups of six, two to a computer, works well. Based on pretests, we grouped students according to their level of readiness with mathematical computations.

ASSESSING STUDENT WORK

Assessing stations can be tricky, especially when differentiating subject matter. When we designed these activities, we decided that we would give percentage grades. Each assignment was graded and the percentages were used to record a letter grade. Safety and work ethic were also included in the grade. Consider how you would grade these assignments individually.

CONCLUSION

Once you have set up stations, you may find that they work for many activities. I have used the same basic stations from the Matter in Motion unit when working with density, as well as atoms. In the atom unit, I substituted building models at the laboratory station.

If all of this sounds daunting, try this quick and easy rotation: Group student desks into pods and arrange them in a row across the room. For a class of 30, two sets of four pods, with four students at each station, works well. Place an instrument with an object to measure at each group of desks. Give students a worksheet so they can record their measurements. Groups spend 20 minutes at each station. This is a quick and easy version of the Station Approach.

Using stations allows you to teach with limited resources, adds variety to your teaching, and lets you do all those lessons you know will help students learn and be successful.

ACKNOWLEDGMENTS

I would like to thank Sarah Harashe, whose organized and forthright style helped make this idea a reality. I would also like to thank Robin Jaques and Michael Jones, who made sure this article made sense.

Resources
Tomlinson, C. 1999. *The differentiated classroom: Responding to the needs of all learners.* Alexandria, VA: ASCD.
Winebrenner, S., and P. Espeland. 2000. *Teaching the gifted student in the regular classroom: Strategies and techniques every teacher can use to meet the academic needs of the gifted and talented.* Minneapolis, MN: Free Spirit Press.

This article first appeared in the February 2007 issue of Science Scope.

Chapter 16

Examining Current Events in Science, Mathematics, and Technology

by John Eichinger

OVERVIEW

The national standards in science and mathematics call for these subjects to be taught from personal and social perspectives, thus strengthening students' decision-making skills. Preeminent science educator Paul DeHart Hurd called for "a curriculum that relates science to human affairs, the quality of life, and social progress" (1994, p. 109). In this activity students examine news coverage not only from the perspective of science, technology, and math, but also based on the story's impact on real people, that is, implications for human rights and social justice. Interdisciplinary connections are embedded in an engaging, accessible, and human context, as students read, analyze, and openly discuss a teacher-selected news article. By facilitating honest dialogue, the teacher helps students confidently face controversial topics and develop crucial critical-thinking skills.

PROCESSES/SKILLS

- Describing
- Analyzing
- Concluding
- Inferring
- Inquiring
- Communicating

RECOMMENDED FOR

Grades 5–8: Individual, small-group, or whole-class instruction
By choosing news articles appropriate to students' ages and ability levels, this activity can be adjusted for any student in these grades.

TIME REQUIRED

1–2 hours

Further Reading

- "Up-to-the-Minute Meteorology," from the February 2004 issue of *Science Scope*

MATERIALS REQUIRED FOR MAIN ACTIVITY

- Enough photocopies of a news article for the entire class (consider newspaper, magazine, and internet sources)

CONNECTING TO THE STANDARDS

NSES
Grade 5-8 Content Standards:
Standard A: Science as Inquiry
- Abilities necessary to do scientific inquiry (especially using appropriate tools to gather data, thinking critically, and considering alternative explanations)

Standard E: Science and Technology
- Understandings about science and technology (especially that perfectly designed technological solutions do not exist)

Standard G: History and Nature of Science
- Nature of science (especially that thorough evaluation and interpretation of investigations is a crucial part of scientific inquiry)

NCTM
Standards for Grades 3-8:
- Communication (especially analyzing and evaluating the mathematical thinking of others)
- Connections (especially recognizing the connections among mathematical ideas and to investigations outside mathematics)

SAFETY CONSIDERATIONS

Basic classroom safety practices apply. If students use online sources, be certain to monitor student web use to avoid contact with inappropriate sites and information.

ACTIVITY OBJECTIVES

In this activity, students

- read and analyze a current event not only for its content in science, technology, or math, but also for its human impact, including human rights and social justice implications.

MAIN ACTIVITY, STEP-BY-STEP PROCEDURES

1. Begin by choosing a current event article from a newspaper, news magazine, or the internet.

The article should be directly relevant to some aspect of science, technology, or math. Because real-world issues seldom fall conveniently under a single subject heading, your article is likely to have indirect connections to other fields. Your choice of current events could raise issues and questions related to history, sociology, psychology, or politics. Be sure to exercise sensitivity to school district policies and community perspectives when choosing a news item. As you make your choice of articles, you might also consider the human rights issues associated with the news event. Such issues are not beyond the scope of the elementary or middle school classroom and are, in fact, highly motivating for students due to the relevance of the topics and the opportunities for authentic dialogue. Take into account the human rights issues associated with news stories regarding global climate change, immunization, cloning, colonization of other planets, organ transplants, environmental hazards, health care, or waste management. An integrated analysis of the news article, including consideration of human rights issues, is promoted by Activity Sheet 1, page 78.

2. Photocopy the article for all class members and read it together, clarifying new concepts and terms as necessary. Have students break into small groups for analysis of the article, with each individual student recording responses on Activity Sheet 1. Facilitate the analysis by moving around the room from group to group, listening, asking, and assessing.

3. Resume whole-class instruction by discussing the groups' results and reactions to the article. Throughout the analysis and discussion, prompt students to notice and express their personal responses to the article. Encourage an awareness and use of authentic student voice, keeping in mind that this activity is designed to illuminate student perspectives via intellectual exploration, not simply to generate standardized, right/wrong responses. Personalize the discussion, especially at the elementary level; for example, ask, "How might a young person like you react to these conditions?" Ask students to consider the article's impact on various demographic groups.

A basic approach to this analysis and discussion is as follows:

a. Clarify the problem. What is going on? Broaden students' understanding of the situation.

b. Define the basic pro and con reactions to the article, concentrating on science, math, and technology connections.
c. Consider the human rights implications: violations, infringements, advancements. Who is affected by the situation, and how are they affected?
d. Through open dialogue, determine workable solutions to the problem. Determine areas of impasse.
e. What must be done to implement the solution(s)?
f. What additional information is needed to help solve the problem?

The teacher has a number of responsibilities in this activity: to help students understand that every problem may not have a simple answer; to help students learn to accept an element of uncertainty; to seek fairness in presenting and discussing the topic; to avoid proselytization and the tendency to oversimplify complex topics; and ultimately to induce authentic, critical thought.

DISCUSSION QUESTIONS

Ask students the following:

1. Do all situations in real life have simple solutions? Explain your answer.
2. When faced with a complex problem, is it a good idea to consider more than one perspective before making any decisions? Explain your answer.
3. What types of careers might involve solving complex problems?

ASSESSMENT

Suggestions for specific ways to assess student understanding are provided in parentheses.

1. Were students able to summarize the chosen article? (Use student responses to Activity Sheet 1 as performance assessment and observations made during Procedure 3 as embedded evidence.)
2. Could students explain the importance of the article in terms of its science, technology, or math

Sample Rubric Using These Assessment Options

	Achievement Level		
	Developing 1	Proficient 2	Exemplary 3
Were students able to summarize the chosen article?	Attempted unsuccessfully to summarize the article	Summarized the article in a general way	Successfully summarized the article, including details and varying viewpoints
Could students explain the importance of the article in terms of its science, technology, or math (STEM) content?	Attempted unsuccessfully to explain the importance of STEM content	Generally explained the importance of STEM content	Explained the importance of STEM content in detail, noting interdisciplinary connections
Were students able to discuss the human rights aspects of the current event?	Attempted unsuccessfully to consider the article's human rights aspects	Generally considered the article's human rights aspects	Discussed the article's human rights aspects in detail, including viewpoints of different people
Did students, through open dialogue, arrive at solutions to the problem, or could they explain why a solution is not yet feasible?	Attempted unsuccessfully to arrive at a solution to the problem presented	Successfully described a solution to the problem presented	Successfully described several solutions to the problem or explained why solutions are not yet fully feasible

Activity Sheet 1

Examining current events in science, technology, and mathematics

Respond to the following based on the news article from class.

1. Summarize the article in 30 to 50 words. List any new words that you don't understand.

2. What does the article have to do with science, technology, or math?

3. Who might be affected by the situation or problem reported in the article? How might they be affected?

4. Why is the article important? (Consider the viewpoints of several different people.)

5. What additional information is needed to resolve the problem reported in the article?

content? (Use student responses to Activity Sheet 1 as performance assessment and observations made during Procedure 3 as embedded evidence.)

3. Were students able to discuss the human rights aspects of the current event? (Use observations made during Procedure 3 as embedded evidence.)

4. Did students, through open dialogue, arrive at solutions to the problem, or could they explain why a solution is not yet feasible? (Use observations made during Procedure 3 and student responses to the Discussion Questions as embedded evidence.)

OTHER OPTIONS AND EXTENSIONS

1. Students, either individually or in groups, might wish to expand their knowledge about the news topic. Encourage them to present their research to the class in the form of a debate, play, poem, video, or art project.

2. Have students write letters related to the news report. They should address the letters to parties in or related to the current event article *and actually send them.* Be judicious about sharing your own perspective so that your students will more readily develop and record their own views. This exercise is especially empowering when the news issue is local and students can see the results of their correspondence.

3. Have students explore news sources for relevant articles of their own choosing. Let them present and discuss those articles in groups or in a classwide forum.

Reference

Hurd, P. D. 1994. New minds for a new age: Prologue to modernizing the science curriculum. *Science Education* 78 (1): 103–116.

Resources

Jennings, T. E., and J. Eichinger. 1999. Science education and human rights: Explorations into critical social consciousness and postmodern science instruction. *International Journal of Educational Reform* 8 (1): 37–44.

LeBeau, S. 1997. Newspaper mathematics. *Teaching Children Mathematics* 3 (5): 240–241.

McLaren, P. 1995. *Critical pedagogy and predatory culture: Oppositional politics in a postmodern era.* New York: Routledge.

O'Connell, S. R. 1995. Newspapers: Connecting the mathematics classroom to the world. *Teaching Children Mathematics* 1 (5): 268–274.

Silbey, R. 1999. What is in the daily news? *Teaching Children Mathematics* 5 (7): 390–394.

This chapter first appeared in Activities Linking Science With Math, 5–8 *(2009), by John Eichinger.*

PART

4

Instructional Lessons That Maximize the Science Budget

Chapter 17

Sun-Savvy Students

Free Teaching Resources From EPA's SunWise Program

by Luke Hall-Jordan

With summer in full swing and the Sun naturally on our minds, what better time to take advantage of a host of free materials provided by the U.S. Environmental Protection Agency's (EPA) SunWise program. SunWise aims to teach students and teachers about the stratospheric ozone layer, ultraviolet (UV) radiation, and how to be safe while in the Sun. Through its website and activity kit, SunWise offers numerous resources to help you add some sizzle to your solar science curriculum—while teaching sun-safe practices students can use their whole lives. Best of all, the resources are easy to get, easy to use, and FREE!

Further Reading

- "More Than Meets the Eye," from the December 2009 issue of *Science and Children*
- "Sunshine on My Shoulders," from *More Picture-Perfect Science Lessons* (2007)

SUNWISE OFFERINGS

Website

When starting off a solar unit, it's a good idea to begin by introducing the basics. On the SunWise website *(www.epa.gov/sunwise)*, you can find a ready-made, grade-appropriate PowerPoint presentation explaining basic solar content in a fun, accurate, and engaging way. The presentations come in three versions for students in grades K–2, 3–5, or 6–8, but they each cover information about what's good and bad about the Sun and UV radiation, and include fun true/false and "What SunWise animal am I?" questions. (Q: "In water and mud I love to stay. My body makes an oily pink sunscreen to protect my skin so I can play! What animal am I?" A: A hippo!) The website also provides downloadable video clips highlighting certain aspects of the Sun's impact on health.

Tool Kit

The tool kit is the core of the SunWise program. Each kit contains more than 50 standards-based activities for elementary and middle school students—including everything you'll need to get started: background information on the science, the health effects, and how you can prevent overexposure. Student worksheets and reproducibles are also part of the tool kit. The most popular part of the tool kit is the SunWise Frisbee, which turns purple when exposed to UV light—even on a cloudy day! The Frisbee can be used to show students the effects of using sunscreens with different sun protection factors (SPFs) or the difference between red and blue cloth's ability to block UV. (I'm sure it comes as no surprise to you that blue is by far the better color choice because it is closer to ultraviolet on the electromagnetic spectrum!)

In addition to being free and fun, using the SunWise program for just one to two hours a year gets results. Through numerous evaluations of the program, we've found that students educated using the SunWise tool kit had an 11% reduction in sunburns, an increase in Sun safety and UV knowledge, a decrease in students' perception that a tan is healthy, and stronger intentions to play in the shade and use sunscreen (Geller et al. 2001; Geller et al. 2003a; Geller et al. 2003b; Kyle et al. 2008). Additionally, teacher survey results show that three out of four teachers using SunWise change their own behavior after using the program.

SAMPLE ACTIVITY

Be a SunWise Traveler

The expanded SunWise Traveler activity is the latest addition to the tool kit. This activity integrates science, computer, math, and social studies skills as students research how people across the globe protect themselves from the Sun's UV rays. The activity helps students make connections and comparisons between their local environment and sun-safe behaviors they should practice when visiting other parts of the world.

For this activity for grades 3–5 (a 6–8 version is available in the kit for older or more advanced students), you'll need 45–60 minutes and the following supplies:

- Maps of the United States and the world
- Computers with internet access
- SunWise Action Steps (Figure 1)

Teacher Background

Many of us may vacation in locations with extreme UV intensity, especially in comparison to the UV intensity at that time of year in our home cities and towns. We don't always realize how intense the Sun is, so unfortunately we don't always adequately prepare for the UV radiation that we are exposed to, resulting in severe sunburns. Ouch! One study found that 88% of sunburns in children occur during holiday or vacation periods (Autier et al. 1998). This can pose a serious potential problem when you combine this

Figure 1

SunWise Action Steps
(from *www.epa.gov/sunwise/actionsteps.html*)

Do Not Burn
Five or more sunburns doubles your risk of developing skin cancer.

Avoid Sun Tanning and Tanning Beds
UV light from tanning beds and the Sun causes skin cancer and wrinkling. If you want to look like you've been in the Sun, consider using a sunless self-tanning product, but continue to use sunscreen with it.

Generously Apply Sunscreen
Generously apply sunscreen to all exposed skin using a Sun Protection Factor (SPF) of at least 15 that provides broad-spectrum protection from both ultraviolet A (UVA) and ultraviolet B (UVB) rays. Reapply every two hours, even on cloudy days, and after swimming or sweating.

Wear Protective Clothing
Wear protective clothing, such as a long-sleeve shirt, pants, a wide-brimmed hat, and sunglasses, where possible.

Seek Shade
Seek shade when appropriate, remembering that the Sun's UV rays are strongest between 10 a.m. and 4 p.m. Remember the shadow rule when in the Sun: Watch your shadow. No shadow, seek shade!

Use Extra Caution Near Water, Snow, and Sand
Water, snow, and sand reflect the damaging rays of the Sun, which can increase your chance of sunburn.

Watch for the UV Index
The UV Index provides important information to help you plan your outdoor activities in ways that prevent overexposure to the Sun. Developed by the National Weather Service (NWS) and EPA, the UV Index is issued daily in selected cities across the United States.

Get Vitamin D Safely
Get vitamin D safely through a diet that includes vitamin supplements and foods fortified with vitamin D. Don't seek the Sun.

Early detection of melanoma can save your life. Carefully examine __all__ of your skin once a month. A new or changing mole in an adult should be evaluated by a dermatologist.

information with the fact that sunburn is a risk factor for skin cancer.

In addition, it is important to understand that UV rays are reflected by snow, sand, water, and pavement. Something to consider at higher altitudes and latitudes is that fresh snow reflects UV radiation. Sand and water also reflect UV radiation and can increase UV exposure at the beach.

The closer you get to the equator, the more intense the UV rays. This occurs because the Sun is more directly overhead, causing a shorter distance for the Sun's rays to travel through the atmosphere, and there is naturally less ozone in the stratosphere in the tropics. The higher in altitude you go, the more intense the UV rays become because there is less atmosphere for the UV to travel through.

So putting all this knowledge together through this activity, we can raise our awareness of the dangers specifically associated with travel to UV-intense destinations—and with this knowledge, hopefully decrease our number of sunburns!

Student Exploration

To begin the activity, engage students by asking them to think about a place to which they would like to travel someday. Or ask students if they have a friend or relative who lives far away and whom they might like to visit. Have students identify the place they would like to visit along with the time of year they would like to do this traveling.

Next have students, working individually or in pairs, use the SunWise website materials (see Internet Resources) to identify the UV Index mean (average) for both where they live and the place they would like to visit and then compare the two locations. For example, a student visiting Denver, Colorado, and living in Boston, Massachusetts, might comment, "On average, the UV Index in Denver in July is Extreme at 11 or higher, while the UV Index in Boston is Moderate to High at 5 to 6. It looks like Denver's high altitude makes the Sun's UV rays more intense."

Ask the children to offer ideas as to why the UV Index is higher or lower at their travel destination. Then have students identify and record the SunWise action steps (see Figure 1) necessary to take when visiting their destination of choice. Going back to the example above, with the UV Index at Extreme in Denver during the month of July, students should record all of the SunWise action steps, including wearing sunscreen, a hat, sunglasses, and full-length clothing and seeking shade during the midday hours. In most instances this will be the case unless the UV Index is below 3. After-

Connecting to the Standards

This article relates to the following *National Science Education Standards* (NRC 1996).

Content Standards

Standard F: Science in Personal and Social Perspectives
- Personal health (K–8)

ward, share their findings as a class. Answers will vary based on students' location choice.

For a modification, you could also have students record the UV Index at your school once a day for two to four weeks near noontime using a UV meter or the forecast for your city or town (see Internet Resources). Students could calculate the average for that time period and compare this number to the average calculated for a place they'd like to visit.

FINAL THOUGHTS

However you choose to use the materials, when you teach students to be sun savvy, not only are you teaching valuable science content but you're also helping instill everyday healthy habits that will benefit students for a lifetime.

References

Autier P., et al. 1998. Sunscreen use, wearing clothes, and number of nevi in 6- to 7-year-old European children. *Journal National Cancer Institute* 90: 1873–80.

Geller, A. C., et al. 2001. The Environmental Protection Agency's national SunWise school program: Sun protection education in U.S. schools (1999–2000). *Journal of the American Academy of Dermatology* 46 (5): 683–689.

Geller, A., et al. 2003a. Evaluation of the SunWise school program. *Journal of School Nursing* 19 (2): 93–99.

Geller, A. C., et al. 2003b. Can an hour or two of sun protection education keep the sunburn away? Evaluation of the SunWise school program. *Environmental Health: A Global Access Source* 2 (13). *http://ehjournal.net/content/2/1/13.*

Kyle, J. W., et al. 2008. Economic evaluation of the U.S. Environmental Protection Agency's SunWise program: Sun protection education for young children. *Pediatrics* 121 (5). From *http://pediatrics.aappublications.org.*

National Research Council (NRC). 1996. *National science education standards.* Washington, DC: National Academies Press.

Internet Resources

Average Monthly UV Index Maps
  www.epa.gov/sunwise/uvimonth.html

Centers for Disease Control and Prevention: Skin Cancer
  www.cdc.gov/cancer/skin
Daily UV Index Forecast
  www.epa.gov/sunwise/uvindex.html
Environmental Protection Agency SunWise Program
  www.epa.gov/sunwise
SHADE Foundation of America
  www.shadefoundation.org/schools.php

Intellicast Weather Forecast
  www.intellicast.com
Weatherbase
  www.weatherbase.com

This article first appeared in the Summer 2008 issue of Science and Children.

Chapter 18

Layered Liquids

by John Eichinger

OVERVIEW

This activity involves an exploration of density. Why does oil float on water? How does drain cleaner sink down the clogged pipe right through standing water? These questions are answered as students make a layered "parfait" of colored liquids based on the varying densities of those liquids. They will calculate densities of the liquid samples as they investigate, describe, and explain the "layered liquids" phenomenon.

PROCESSES/SKILLS

- Observing
- Problem solving
- Predicting
- Describing

Further Reading

- "Shampoo, Soy Sauce, and the Prince's Pendant: Density for Middle-Level Students," from the October 2006 issue of *Science Scope*

- Analyzing
- Concluding
- Measuring
- Calculating
- Inquiring
- Communicating
- Cooperating

RECOMMENDED FOR

Grades 5–8: Small-group instruction
To adapt the lesson for fifth graders, you can measure out the 100 ml samples ahead of time, assist with data collection and analysis, or even undertake the investigation as a whole-class activity.

TIME REQUIRED

1–2 hours

MATERIALS REQUIRED FOR MAIN ACTIVITY

- Water
- Food coloring (one color)
- Corn syrup
- Tall, clear, plastic containers (empty water bottles, approximately 500 ml capacity, are just right)

- Beakers or graduated cylinders for measuring liquids
- Paper cups
- Balances
- Calculators
- Maple syrup
- Vegetable oil
- Dishwashing detergent
- Mineral oil

CONNECTING TO THE STANDARDS

NSES

Grade 5-8 Content Standards:

Standard A: Science as Inquiry
- Abilities necessary to do scientific inquiry (especially observing carefully, thinking critically about evidence to develop and communicate good explanations, and using mathematics effectively)
- Understandings about scientific inquiry (especially recognizing the importance of mathematics in science and noticing that scientific explanations emphasize evidence and logically consistent arguments)

Standard B: Physical Science
- Properties and changes of properties in matter (especially investigating density as a property of matter)

NCTM

Standards for Grades 3-8:
- Numbers and Operations (especially working with numbers and operations to solve problems)
- Measurement (especially understanding and applying the metric system)
- Reasoning and Proof (especially engaging in thinking and reasoning)
- Communication (especially communicating their mathematical thinking clearly)

SAFETY CONSIDERATIONS

Basic classroom safety practices apply. Be sure to demonstrate proper handling of the various liquids.

ACTIVITY OBJECTIVES

In this activity, students

- layer the four liquid samples and explain their results; and

- determine the layered position of a fifth, unknown, liquid based on its density, which they will calculate.

BACKGROUND INFORMATION

Density is the relative weight of an object, defined mathematically as the object's mass divided by volume. A more dense object or material has more tightly packed internal particles. A brick, for instance, is denser (that is, has more tightly packed particles within it) than a piece of wood (whose particles are more loosely packed). A brick is denser than water, and it will sink. Most wood, however, is less dense than water, allowing it to float. Therefore, it is not an object's weight alone that determines whether it will sink or float; it is the object's weight (really its mass) divided by its volume. Consider a large piece of Styrofoam (say, 500 kg): It will float in water despite its large size because it is less dense than the water. That is, it has less mass per unit of volume than water. Or, put another way, if we have two equal volumes (say, 250 cm^3) of Styrofoam and of water, the Styrofoam will be lighter in weight (or contain less mass).

MAIN ACTIVITY, STEP-BY-STEP PROCEDURES

1. Begin this preliminary demonstration by showing the class a two-layered liquid "parfait": water "floating" on corn syrup. (The effect is more dramatic if you first mix a little food coloring into the water and if you let students see you pour the two liquids carefully together.) Ask, "Why do you suppose these liquids form into two layers?" Accept divergent answers, but help students see that density is the reason.

2. Working together in small, cooperative groups, students begin by measuring out 100 ml each of four different liquids: water, maple syrup, vegetable oil, and dishwashing detergent. Samples must be poured into identical containers (because their masses will be compared); paper cups work well. Students should list the substances by sample number in Activity Sheet 1, Table 1, page 89.

3. Next students predict the order of the layers that the four samples will form when poured carefully into the same jar. Which will be on top, and so on? They should record their predictions in Activity Sheet 1, Table 2.

4. Using a balance, students measure the mass of each of the samples. (This is why they need to

be in identical containers; subtract the mass of the cup, weighed when empty, from the mass of each cup when filled with the liquid sample.) Students should record all data in Activity Sheet 1, Table 3.

5. Students should calculate the density of each liquid (mass divided by volume: 100 ml each). Then they should record the densities and again predict the order of the layers that will form when the four samples are poured into the same container (using Activity Sheet 1, Table 4). Ask students, "Did your prediction change? Why or why not?"

6. Students should carefully pour the liquids into a single, tall, clear container, one at a time over a spoon so that they don't mix (see Figure 1). Then they record the results of the layering effect in Activity Sheet 1, Table 5, page 90. Ask students to explain their results, particularly in relation to their predictions.

7. Before having students pour out their samples, try this method of performance assessment or application of the concepts: Each group receives a fifth liquid sample (for instance, mineral oil; again 100 ml in an identical paper cup). Students must determine where in the layered column the liquid will come to rest by measuring its mass,

Figure 1

Pour gently over a spoon to prevent mixing.

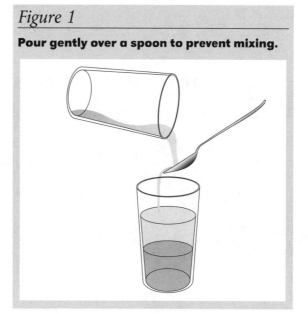

calculating its density, and using the density to predict correctly (in Activity Sheet 1, Table 6). Then students can pour the fifth liquid into the column to check their calculation, analysis, and prediction. They can write science journal entries about how density affects floating and sinking.

Sample Rubric Using These Assessment Options

	Achievement Level		
	Developing 1	Proficient 2	Exemplary 3
Were students able to successfully layer the liquid samples?	Attempted to layer but were unsuccessful	Successfully measured and layered their liquid samples	Successfully measured and layered their liquid samples and took a leadership role in data collection and analysis
Were student predictions correct? If so, could students explain why? If not, could they explain why not?	Attempted to explain their predictions and results but were not able to do so to any significant extent	Effectively explained their predictions, whether accurate or not, in terms of their data and results	Effectively explained their predictions, whether accurate or not, in terms of their data and results, and used math concepts as part of their explanation
Could students successfully determine the layered position of the fifth liquid, and did they explain how they arrived at their answers?	Attempted to predict the position of the unknown liquid but were not successful	Successfully predicted the position of the unknown liquid	Successfully predicted the position of the unknown liquid and were able to explain their rationales using math concepts and data examples

DISCUSSION QUESTIONS

Ask students the following:

1. What does density have to do with sinking and floating?
2. If you tried the layering activity aboard the space shuttle in outer space, would the results differ, and if so, how? (The liquids are weightless, so no layers form.) What if you tried the activity on the surface of the Moon? (The liquids would have the same layers as on Earth, despite lower gravitational pull.)
3. Can you think of any jobs that might involve the concepts of density, sinking, or floating? Explain your answer for each job that you can name.

ASSESSMENT

Suggestions for specific ways to assess student understanding are provided in parentheses.

1. Were students able to successfully layer the liquid samples? (Use observations made during Procedures 2–6 as performance assessments.)
2. Were student predictions correct? If so, could students explain why? If not, could they explain why not? (Use Activity Sheet 1 as performance assessment, and use responses to Discussion Questions 1–3 as embedded evidence or as writing prompts for science journal entries.)
3. Could students successfully determine the layered position of the fifth liquid, and did they explain how they arrived at their answers? (Use observations made during Procedure 7 as performance assessment, and use student analysis of that procedure as a prompt for a science journal entry.)

OTHER OPTIONS AND EXTENSIONS

1. Groups can place some small objects (pieces of wood, cork, rock, eraser, wax, fruit, plastic, metal, etc.) into the layered column (from Procedure 6), predicting where each will come to rest. Students will see that density applies to solid as well as liquid matter.
2. Try the basic activity using water and any sort of fruit juice, soda, or other water-based mixture. The distinct layers will not form quickly, if at all (you may need to let the final product settle for several hours before any layering is evident). Ask students to explain why the results differ so significantly from the basic activity. (The fruit juice or soda is composed almost solely of water, so it is neither more nor less dense than the pure water sample. The two samples will mix, and clear layering based on density is not evident.)

Resources

Beckstead, L. 2008. Science journals: A creative assessment tool. *Science and Children* 46 (3): 22–26.

Bricker, P. 2007. Reinvigorating science journals. *Science and Children* 45 (3): 24–29.

Halpin, M. J., and J. C. Swab. 1990. It's the real thing—The scientific method. *Science and Children* 27 (7): 30–31.

Nesin, G., and L. Barrow. 1984. Density in liquids. *Science and Children* 21 (7): 28–30.

Pearlman, S., and K. Pericak-Spector. 1994. A series of seriation activities. *Science and Children* 31 (4): 37–39.

Scheckel, L. 1993. How to make density float. *Science and Children* 31 (3): 30–33.

This chapter first appeared in Activities Linking Science With Math, 5–8 *(2009), by John Eichinger.*

Activity Sheet 1

Layered liquids

Table 1

	Sample	Which Substance?
1		
2		
3		
4		

Prediction 1: In what order will the sample layers end up?

Table 2

Layer Order	Sample	Substance
Top Layer		
Second Layer		
Third Layer		
Bottom Layer		

Table 3

Sample	Substance	Mass (g)	Volume	Mass/Volume	Density
1			100 ml	/100 ml	
2			100 ml	/100 ml	
3			100 ml	/100 ml	
4			100 ml	/100 ml	

Prediction 2: After calculating the densities of the samples, what order do you think the final layers will take?

Table 4

Layer Order	Sample	Substance
Top Layer		
Second Layer		
Third Layer		
Bottom Layer		

Results:

Table 5

Layer Order	Sample	Substance
Top Layer		
Second Layer		
Third Layer		
Bottom Layer		

Explain your results:

Data for Sample 5:

Table 6

Sample	Substance	Mass (g)	Volume	Mass/Volume	Density
5			100 ml	/100 ml	

Prediction: Where will the Sample 5 layer be in relation to the other four samples, and how do you know?

Results, using Sample 5:

Chapter 19

Button Basics

Prompting Discussions of Properties

by Sarah J. Carrier and Annie B. Thomas

Elementary teachers of science are at a great advantage because observation—collecting information about the world using our five senses—and classification—sorting things by properties—come so naturally to children. Many examples of classification occur in science: Scientists, for example, group things starting with large categories, such as living or nonliving, and then further classify them into more distinct groups based on specific properties. Living things could be plants or animals, and animals could be vertebrates or invertebrates. Vertebrates could then be further classified as birds, reptiles, amphibians, mammals, or fish, etc., with further classifications made by paying closer and closer attention to more specific, shared properties. Classification is always based on what the scientist is studying—those same animals, for example, could be classified differently, into such categories as omnivore, carnivore, herbivore, and so on.

Buttons are ideal objects when teaching children about properties and classification. These familiar and inexpensive objects provide a meaningful teaching tool in the classroom. This lesson can be presented either as a basic introduction to classification by properties, or it can be used to offer increasingly complex explorations that help students experience a more in-depth understanding of properties. We used the following lesson with fourth-grade students, but the lesson can

easily be adapted for use with children at any cognitive and developmental level or grade.

GETTING STARTED

The only materials needed for this lesson are a bag full of buttons and premade Venn diagram mats (described on p. 92). We purchased various boxes of buttons at a discount store with little expense. The only requirement is that the collection contains buttons with various characteristics—different colors, sizes, textures, number of holes, etc. Enough buttons to fill half of a gallon-size plastic storage bag are sufficient for most classrooms.

We began the activity with a discussion of the word *property*. Children were quick to provide their own definitions, such as "characteristics," "features," and "qualities." One student assumed the word meant "land." After a discussion of multiple-meaning words and context clues, we related the different meanings of *properties* depending on the usage. To illustrate properties further, we pointed out the children's different characteristics. We spent approximately 10 minutes sorting ourselves by eye color, hair color, type of shoes, height, and number of siblings. Modeling similarities and differences in this way set a basic foundation to compare buttons and then ultimately analyze them in a Venn diagram.

BRING OUT THE BUTTONS

As we walked around the room to distribute a big handful of buttons to each predetermined group of three to four students, we reminded them not to put the buttons into their mouths—sense of *taste* would not be needed here! However, sight, touch, and maybe even smell or hearing would be involved. (For example, students can try to sense odor in wooden buttons; the sound of buttons dropping varies depending on the size, material, and button surface.)

Once everyone had buttons, we explained the next task: Observe the similarities and differences among the buttons and discuss these findings with your group. As we walked around the room, students discussed grouping buttons by different characteristics or properties. Many started with colors or solids compared to patterns. Children were excited to share their own observations. We heard such comments as "This one still has string on it!" "What do you call this thing on the back of the button with no holes?" and "My favorite button is this one!" (Many students quickly identified their "favorite" buttons.)

After a few minutes, we called the class to attention to discuss their observations. Each group explained the various properties of their own set of buttons. Most groups included color, size, shape, pattern, texture—whether the buttons were rough, smooth, scratchy, or bumpy—and number of holes or shank (those without holes). Some groups even created additional relevant properties we had not considered, such as "shades of red" category, which included red buttons and various shades of pink.

After listing and recording all of the observed properties on the chalkboard, we asked the students to classify the buttons by properties. As we circulated around the room, we conducted formative assessments to check for understanding of the terms by asking the groups what characteristics or properties they used to form their groups and to explain their sorting choices.

To challenge the groups, we walked to each desk and removed all except one of each group's piles of buttons. An example of this would be leaving only the pile of blue buttons from the groups of buttons classified by color. We then instructed them further: "Now sort *this* group using another property." After a few moments of questioning facial expressions, most students realized that they could further sort these blue buttons by another property, such as size.

VENN DIAGRAMS FOR CHILDREN

A Venn diagram mat is simple to make: We drew two overlapping circles on a piece of 12 in. × 18 in. construction paper and laminated it. While standing in front of the class so all could view my simple organizational tool, we explained how Venn diagrams would help us sort things by characteristics.

Students had previously practiced using this diagram with a simple activity using pattern blocks to manipulate and visualize organization by properties. We had labeled one circle "Blue" and the other circle "Square." Students quickly understood that blue squares belonged in the overlapping section that was part of both the blue and the square circles. Various other colored squares belonged in the non-overlapped square circle, and all of the other blue shapes besides squares belonged in the non-overlapped section of the blue circle.

On the day of the button activity, we reviewed how to do a Venn diagram and modeled a more symbolic diagram, labeling one circle as "Dogs" and the other circle as "Cats." As a class we discussed the similarities and differences and the placement of the characteristics of each. Some examples of shared properties in the overlapping section of this Venn diagram were "mammal," "fur," or "pets." Students told me that properties unique to dogs were "bark" and "wag tail when happy." Properties unique to cats were "climb trees" and "litter boxes." We now felt students were ready to use the diagram with buttons.

SORTING BUTTONS

Each group shared one diagram. We asked the children to choose only two button properties they had discussed and label each circle on their Venn diagrams with one property. Students had to decide on properties that would have some buttons belonging in each circle, some in the overlapping area of the Venn diagram, and some outside the circles. This decision was difficult for students. We challenged them to create their own labels for the circles. For example, if students label one circle for buttons with two holes and the second circle for buttons with four holes, there will be no overlap. This trial-and-error process is valuable for student problem solving and decision making that is very much a part of science. The class spent 15–20 minutes creating their diagrams.

An example of an effective Venn diagram would show one circle labeled for buttons with four holes and the other circle labeled for buttons that are black. The "Four Hole" circle would contain buttons with four holes that are not black. The "Black" circle would contain buttons that are black and have either shanks or two holes. The overlapping middle section would contain buttons that are both black in color and have four holes. The area outside the circles would contain the buttons that have none of the properties of the named circles.

Observing children working with the diagram served as a useful informal summative assessment—we were able to check for student understanding of properties as well as their abilities to classify by similarities and differences. When students had difficulties labeling the circles, we posed questions to clarify or to guide them to choices that allowed for unique button characteristics as well as overlap. For example, one group labeled a circle "Blue" and the other circle "Brown." There was no overlap here, so we asked them to create other properties that are unique but also shared. They agreed to use color for one circle and number of holes for the second circle.

SHARING PROPERTIES

At the conclusion of the activity, each group shared their Venn diagrams, both the ones that had buttons in each of the three areas and the ones that needed revisions. The efforts that required revisions were presented as processes rather than failures. We explained to the class that in science we often come to dead ends that require alterations and shifting approaches. These are not mistakes or failures but rather redirections. We talked about how scientists organize the world through classifications. When a scientist finds an unknown animal,

Connecting to the Standards

This article relates to the following *National Science Education Standards* (NRC 1996).

Content Standards
Standard B: Physical Science
• Properties of objects and materials (K–4)

he or she can determine that it is a vertebrate because it has a backbone. The next question is whether it falls into the category of bird, reptile, amphibian, mammal, or fish, depending on its characteristics.

This button activity introduces and reinforces the meaning of the term *properties* in science. It provides students with experiences using the science-process skills of observing and classifying and allows them to experience the decision-making strategies used in science. Best of all, this activity is easy and affordable.

Reference

National Research Council (NRC). 1996. *National science education standards.* Washington, DC: National Academies Press.

This article first appeared in the January 2008 issue of Science and Children.

Chapter 20

Growth Potential

by Dana M. Barry

Students enjoy carrying out an exciting and challenging research project that combines science with computers and mathematics to investigate how polyacrylate animals change in size over time when placed in water and aqueous salt solutions. The hands-on activity motivates students and provides them with the necessary skills and information to have enjoyable and rewarding science project experiences. Here they have an opportunity to solve a problem and to use the science process skills of observing, collecting, organizing, and analyzing data. Project results are displayed on individual posters, creatively prepared by the students. A schoolwide poster session is a great way to share the information and to build students' confidence and self-esteem.

In this activity, students determine how the percentage of NaCl (table salt) in an aqueous mixture affects the rate of water absorption (growth) by a polyacrylate animal. To start, each student is given a polyacrylate animal. These animals are molded models made of polyacrylate, a hydrophilic substance that attracts water molecules. They cost about $1 each and may be purchased at toy stores or from science supply catalogs. Use a variety of animal types and colors, but be certain to have four identical animals of each type—one for each of the four possible salt solutions in order to eliminate variables. I like to take a photo of each student holding

his or her animal before we begin the project and again at the end so the difference can be observed; however, this is optional. Students then examine their animals, give them a name, and record their observations (length, mass, etc.) on a data table.

Provide each student with 500 ml of an aqueous salt solution (either 0, 1, 2, or 3% salt) in a large labeled container with a cover. Empty plastic food containers from school cafeterias work well. The 0% salt solution is pure tap water. The 1% salt solution has 5 g of salt per 500 ml of solution. The 2% salt solution has 10 g of salt per 500 ml of solution. The 3% salt solution has 15 g of salt per 500 ml of solution. The students are provided with these solutions and told the percentages in advance. Have them put the animals in their containers, cover and label them, and place them in a safe location. Containers may be stored in cabinets or on spacious countertops. For five consecutive days, students take daily measurements of the animals' length and mass, using metric rulers to determine length and a triple-beam balance with a 150 g capacity to measure mass. To do this, students open their containers and remove the animals from the salt solutions. Then they gently pat the animals with paper towels to remove excess water before taking measurements. Each student should wipe off any moisture on the balance after using it. When finished, the animals are put back in the containers, the covers are replaced, and the containers are returned to the storage shelf. After the last set of measurements has been made, take another photo of each student holding his or her animal.

Once the data gathering is complete, students prepare one graph showing elapsed time in hours versus length in centimeters and another graph showing elapsed time in hours versus mass in grams. These are then compared and discussed as a class. Students pair up with other students that have an animal of the same type and color, but of a different percent salt solution. They share data and graphs and compare results, and then draw conclusions about the effect of sodium chloride on the growth of their animals. Select pairs to orally share

information about their animal experiment with the class. The research results indicate that the more concentrated the salt solution, the less the increase of mass and length for each animal. This is because the salt hinders the water's ability to bond with the polyacrylate.

Additionally, students are expected to submit a short report of their research, which could include images and background information collected from the internet and library. Each report should include a description of what was done, the results, and concluding statements. Encourage students to creatively design individualized posters for their research projects. They can use their written reports, photos, data tables, graphs, polyacrylate animal pictures, and background

Amazing Animals Activity

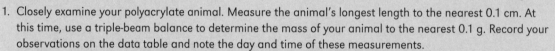

Materials (per student)
- One polyacrylate animal
- Metric ruler
- Triple-beam balance (150 g capacity)
- Assigned aqueous salt solution (either 0, 1, 2, or 3%) in a large plastic container with a top

1. Closely examine your polyacrylate animal. Measure the animal's longest length to the nearest 0.1 cm. At this time, use a triple-beam balance to determine the mass of your animal to the nearest 0.1 g. Record your observations on the data table and note the day and time of these measurements.
2. What do you think will happen to the animals in the different salt solutions?
3. Put your animal in 500 ml of the aqueous salt solution (either 0, 1, 2, or 3% salt as assigned). Write your name, animal type, color, and percent salt solution on the label. Then cover the container and place it in a safe location.
4. For the next four days, make daily measurements of length and mass (at approximately the same time each day). Before making measurements, gently pat the animal with a paper towel to remove excess water. Record your data on the table.
5. After the measurements are complete, prepare two graphs (either by hand or using a computer if available). One graph should include elapsed time in hours versus length in centimeters. The other graph should include elapsed time in hours versus mass in grams. Compare and discuss your results with the class.
6. Conduct research on your experiment and prepare a short report that includes a description of what was done, the results, concluding statements, photos, data table, graphs, animal pictures, and background information about the animal. Design a poster that displays your research to share with the class.

Student's name: _____   Animal color: _____

Animal type: _____   Concentration of the salt solution used: _____

	Time	Elapsed Hours	Length (cm)	Mass (g)
Day 1				
Day 2				
Day 3				
Day 4				
Day 5				

information for the display. As an extension, students can obtain information about the real animals by using a search engine on the internet or encyclopedias in the library, listing at least one reference in their report. Each student should prepare a three-page written report that includes a statement of the problem, a list of materials needed, a procedure to carry out the experiment, concluding statements, and references. The data table and two graphs are used to complement the written report. Students are assigned points for this activity, which culminates in the report and poster display. The project is worth 100 points, or the equivalent of a unit test grade. Students may receive points for including the title page, problem statement, list of materials, procedure, data table, graphs, concluding statements, references, and overall poster display (e.g., appearance and creativity).

This exciting activity provides students with a challenging research project that combines science with computers and mathematics. Students exercise problem-solving and critical-thinking skills as they investigate how polyacrylate animals change in size over time when placed in aqueous salt solutions. They also develop the technical skills of measuring and weighing. In addition, they prepare graphs, reports, and posters, and use the internet as a source of scientific information.

Acknowledgment

This research project was successfully carried out by students in Horizons 2002 and 2003, a special program for gifted seventh- and eighth-grade girls in New York State. It is a component of the Pipeline of Educational Programs offered at Clarkson University in Potsdam, New York.

This article first appeared in the April 2004 issue of Science Scope.

Chapter 21

Precipitation Matters

by Thomas McDuffie

A lthough weather, including its role in the water cycle, is part of most elementary science programs, an in-depth examination of raindrops and snowflakes is rare. Together rain and snow make up most of the precipitation that replenishes Earth's life-sustaining freshwater supply. When viewed individually, raindrops and snowflakes are quite varied either in size or shape and provide surprising hints about the atmospheric conditions in which they formed.

The activities described in this article invite children in grades three through six to study the sizes of raindrops and the shapes of snowflakes as an extension of a weather or water unit. In the process, myths about the tear shape of raindrops and the standard shape of snowflakes should be dispelled. Students learn that snowflakes and raindrops are more complex than greeting cards suggest.

Further Reading

*   "Snowflake Symmetry," from the December 2009 issue of *Science and Children*

These activities can be taught regardless of local weather conditions or time of the school year: Students observe "virtual" snowflakes using web-based resources. Raindrops are "captured" in flour and then examined to approximate the height from which they fell.

BACKGROUND

Rain and snow, the planet's principal sources of freshwater, can be traced through the hydrologic or water cycle—evaporation, condensation, and precipitation. The cycle's processes occur simultaneously and continuously (see McDuffie and Palmer 2000 for a fuller explanation).

During condensation, water vapor (a gas) becomes a liquid or solid. For this transition to occur, water vapor must saturate the air, that is, the relative humidity must be at least 100%. Tiny particles of dust, smoke, or salt that attract water molecules provide a physical center or seed for the condensation process. Outside the tropics, condensation generally occurs high in the atmosphere where below-freezing temperatures produce ice crystals that fall, melt, and join together to form droplets. When water droplets become large and numerous enough, they form clouds.

Precipitation in any form—rain, sleet, hail, or snow—involves the formation of droplets or particles that grow through a continuous process of collisions and re-evaporations within clouds; large drops collide with and absorb smaller ones. When ice crystals or raindrops have floated, bounced, and combined long enough and grown massive enough to overcome the upward movement of air, gravity pulls them to Earth. This means that drizzle or gentle rains on calm days consist of small drops formed relatively near the ground. On windy days when cumulus clouds fill the sky, drops can bounce around much farther from the ground. This means that the size of raindrops increases as the height where they formed increases. Consequently, the huge drops formed in summer thunderstorms began high in the atmosphere as ice crystals that grew for many minutes—even hours—before melting on the way down. Conversely, the very large snowflakes that float down like fall leaves result from collisions when temperatures are close to 0°C.

COLLECTING RAINDROPS

Our procedures for capturing raindrops are borrowed from naturalist Wilson A. Bentley, better known for his photographic record of snowflakes. Learning about Bentley's life through websites or the book *Snowflake Bentley* (Martin 1998) adds a human dimension to the pursuit of science. Bentley studied raindrops using an ingeniously simple method to capture raindrops to compare the sizes and shapes in different storms.

He allowed raindrops to fall on finely ground flour to form dough balls. The "drops" can be separated from the loose flour using a mesh strainer. After drying for a couple of minutes, the dough balls can be used immediately or stored in a freezer in dated plastic bags. If at all possible, children should collect drops from at least one rain event in a location away from buildings or trees. Rather than hope a rain dance works, however, teachers should collect and refrigerate some "floured raindrops" from three or four rains before introducing the unit. (I admit to first testing the procedure under the shower!)

ACTIVITY 1: INVESTIGATING RAINDROPS

Purpose
To observe the size and shape of raindrops and to estimate how far they fell

Background
When rain falls onto flour, it makes dough balls the same size as the drop itself. This approach was used by Wilson Bentley almost 100 years ago to study rainstorms. He found out that the size of drops varied

depending on the type of rainstorm. Gentle rains had small drops, while thunderstorms were made up of very large drops. Scientists now know that large drops fall from greater heights than small ones. As the diameter of drops increases, the volume of water they contain grows even faster. If one drop has twice the diameter of another, it contains eight times as much water.

Materials
- For the class: 1 lb. bag powdered flour, resealable plastic bags, and a fine mesh strainer (plus teacher-collected samples from prior rain events)
- For each group of three or four students: one 6 in. to 8 in. aluminum pie pan and a piece of cardboard to cover it, ruler, graph paper, dark construction paper, and a sharpened pencil

Procedure
1. Sift 1–1.5 cm of flour directly into each group's pie pan using the strainer. (Doing this over a wastebasket or cardboard box reduces mess.)
2. Practice uncovering and covering the flour with the cardboard before going outside; proceed to the collecting site, collect raindrops for 8–10 seconds, and then return to the class. (I pair up children to share umbrellas if anyone lacks rain gear.)
3. Separate the dough balls over a wastebasket or box using the strainer.
4. Dump the "raindrops" onto colored construction paper and allow them to dry for about two minutes.
5. Teams of two or three should observe, then line up 10 dough balls and measure the length of each one. Next, they should determine the average diameter and record this result.
6. Sketch the drops and compare the shapes of small, medium, and large drops. Arrange the drops from smallest to largest. (It's easiest to sort along the edge of a ruler, moving the dough balls carefully with a pencil.)
7. Using drops from several rain events, compare drop sizes between storms.

Questions
- Describe the shape(s) of the drops. *(Round or slightly egg shaped)*
- Are the sizes of the drops the same or different? *(Storms usually include one to three groups of drops with the same size.)*
- Are there more large, medium, or small drops? *(Depends on the storm)*
- Draw a bar graph relating the size of drops to their number.

Figure 1

Virtual snowflake samples

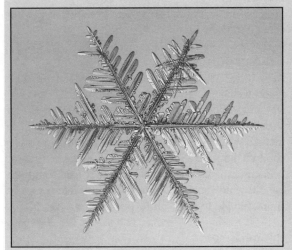
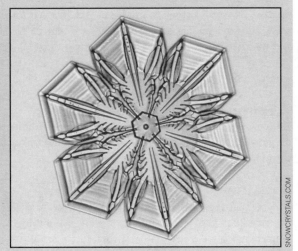

- Do you think most of the water in your rainstorm comes from the large or the small drips? Why? *(Predictions are based on the number and volume of larger versus smaller drops. No exact answer is expected.)*
- Do all storms have the same drop sizes? *(No, there is a range of drop sizes.)*
- Did all drops in a rainstorm fall from the same height? *(If they are the same size, yes. If sizes differ, they fell from different heights.)*
- What happens to rain that falls on your schoolyard? *(Some goes into the ground and some runs off.)*
- How is this related to the water cycle? *(It becomes part of the groundwater or goes into a stream.)*

Follow-Up Discussion

Children should learn several things as a result of the activity: Raindrops have different sizes but the same shape; their size indicates how far they fell; and large drops contain more water than small ones. The surprise evident in children's reactions to the formation of dough balls and their varied sizes is a good starting point for a discussion. Invariably someone asks, "What causes the balls to form?" Classmates can than share ideas—the water is trapped in the flour, the flour absorbs the water, and it's like cement. The size and shape of the dough balls are related to the raindrops that formed them. After the relationship between the size of raindrops and height and altitude has been shared by the teacher, youngsters can predict which drops were formed at higher altitudes. Visually

comparing the space taken up by 10 marbles and 10 golf balls is a good analogy for the volume of rain associated with large versus small drops. No follow-up is complete until raindrops, rain, and precipitation are related to the water cycle. That is, students realize that most of the Earth's freshwater falls as rain, which can be traced through this cycle. Younger children can trace the flow of water through a watershed to a lake or the ocean. More sophisticated learners can extend this to such topics as evaporation from the land and transpiration from plants, movement of groundwater, and water being trapped in glaciers.

EXAMINING SNOWFLAKES

If your class is lucky enough to experience a snow event when studying precipitation, it's magical for a child to capture a few flakes on a mitten and examine them with a magnifying glass. What did they see? Did anyone find a perfect six-sided snowflake? It's not easy! They occur less than 25% of the time and less often still when the temperature is near 0°C. Ice crystals usually break when they bump into one another or when they melt and refreeze. Few crystals survive the collisions with the perfect six-sided symmetry found on holiday greeting cards.

While the firsthand experience is a magical way to study snowflakes, timing and sometimes location make it basically impossible. So I searched for alternative ways of preserving the flakes. Supposedly clear lacquer or hair spray can be used to "freeze" the shape of snow crystals on cold glass or plastic, but my many attempts over several years involving a wide range of sprays, varnishes, and finishes have been dismal failures. As a result, I turned to images on the internet for a predictable, valuable, and easily managed alternative (see Internet Resources).

ACTIVITY 2: PAPER SNOWFLAKES

Purpose

To show the basic patterns and range of shapes found in snowflakes and to provide a basis for comparison with virtual snow crystals

Background

Making paper snowflakes is a change of pace and the results make wonderful classroom decorations. Numerous patterns are available, but to be scientifically accurate they must be six-sided. Children find it easiest to begin with a circle of white copy paper that can be folded in half, and then the halves folded in thirds.

Materials

- Copy paper, scissors
- Dark construction paper and glue (optional)

Procedure

1. Each student should cut a circle out of the copy paper.
2. Circle should be folded in half, then the half into thirds.
3. Cut the folded paper to form the desired pattern.

Questions

- Write a list of words that describe your flake. *(Pretty, pointy, same number of points or arms, some have more straight sides, some have holes)*
- Compare your flake with others in your group. List similarities and differences. *(More/less pointy, more/fewer holes, more/fewer sides, larger/smaller)*
- Why are the flakes different? Are any in the class exactly the same? *(They were cut differently; some scissors cut better; they made different patterns. They look almost the same; are very similar.)*

So far, we have looked at paper flakes. If the same basic folds were used, each of the six sections is similar and they are flat. Real snowflakes are made up of tiny particles, which must come together in three dimensions. The results can be observed using photographs that magnify actual snowflakes.

ACTIVITY 3: VIRTUAL SNOWFLAKES

Purpose

To show the structure and variety in snowflakes and to classify them by shape

Background

A website dedicated to Bentley's work is a good place to begin. This can be followed by visits to *SnowCrystals.com* and the Buffalo Museum of Science website (see Internet Resources). *SnowCrystals.com* includes a range of information for teachers and older children. Both sites provide galleries of images that can be used for developing a classification scheme for grouping differently shaped crystals. (Note: If internet access is unavailable, images can be printed and copied.)

Procedure

1. Open the Snowflake Bentley website *(http://snowflakebentley.com)*, navigate to the sample snowflakes, and observe the different shapes.
2. Access the images in one of the galleries in the Snow Crystals site (Figure 1, p. 101). *(www.its.caltech.edu/~atomic/snowcrystals)*.
3. Test your classification scheme by applying it to a sample of Bentley's original photos. *(http://bentley.sciencebuff.org/collection.htm)*.

Questions

Using information on the Bentley website, answer the following questions:

- Where did Bentley live? *(Vermont)* What was his nickname? *(Snowflake)* How did the family make its living? *(Dairy farming)*
- How do the images he photographed compare with the paper flakes your group created? *(More shapes and less exact than the paper flakes. Paper flakes look more alike. Paper flakes are pointier. Some snowflakes look prettier. Snow has different shapes.)*

Based on the photos found on *SnowCrystals.com*, sort the images by shape.

- How many groups did you develop? *(Usually three or four, depending on the grade level)*
- What are the outstanding characteristics of each category? *("Look like" stars, cylinders, blobs, etc.)*
- Create a name for each category. *(Stars, pretty, cylinders, tubes, blobs, flowers)*
- Using the groups you identified, classify the Bentley images found at *http://bentley.sciencebuff.org/collection.htm*.

- Do all the flakes fit? (*Kids usually force them to, but some belong to a couple of categories.*)
- How would you modify the scheme to include more flakes? (*Add other categories*)
- If snow fell on your schoolyard, what would happen to the snow that melted? (*Some goes into the ground and some runs off*)
- How would it flow through the water cycle? (*It becomes part of the groundwater or goes into a stream.*)

Follow-Up Discussion

Students readily grasp the idea of different shapes in the paper flakes. The variety in real flakes creates a challenge. At the most basic level, photos of snowflakes are grouped based on visual appeal—"pretty" versus "ugly." The idea of common shapes needs to be teased out. "What does it remind you of?" elicits responses that can be used to establish basic categories—stars, pretty, cylinders, tubes, blobs, flowers.

Inevitably two questions will be asked: "How do snowflakes form?" and "What causes the shapes?" A developmentally appropriate explanation is that tiny ice crystals form, then join together like pieces of a puzzle. When flakes are forming, they often crash into one another and break, so the pattern is changed. Also, air temperature causes partial melting so flakes stick together. As with rain, teachers should relate snow to the water cycle. For example, they might ask, "What happens to snow that falls in the schoolyard?" Just like the rain, it becomes part of the water supply—and evaporates, condenses, and so on.

SHOWERS OF LEARNING

Although no single approach is best for all classes or grades, learning can be assessed authentically through teacher observations and student reports, or it can be limited to the various questions accompanying each activity. Relating precipitation to the larger water cycle helps extend student understanding of the effects of weather.

Teachers have been impressed by the interest these activities generate in a rarely studied aspect of the water cycle. The idea that raindrops can be captured and vary in size is novel. Similarly, the idea that rain falls from different heights escapes most. Yet, it is the large drops

falling from summer thunderstorms that produce crop-damaging and life-threatening flash floods.

The beauty of snowflakes captivates interest. The range of shapes is striking, so the process of classification is a challenge best overcome through discussion and scaffolding. With guidance, students can develop and extend usable classification schemes for snowflakes.

While raindrops and snowflakes are fascinating to study, their role in the water cycle is central. What happens to rain and snow that fall on the school's area? Where does the moisture go? When children begin to explore these questions, their understanding of the interrelationships between runoff, streams, and the journey of water can begin, and precipitation can be considered the starting point of the cycle.

Connecting to the Standards

This article relates to the following *National Science Education Standards* (NRC 1996):

Content Standards
Standard D: Earth and Space Science
- Changes in Earth and sky (K–4)
- Structure of the Earth system (K–4)
- Properties of Earth materials (5–8)

References

Martin, J. B. 1998. *Snowflake Bentley*. Boston: Houghton Mifflin.

McDuffie, T. E., Jr., and A. Palmer. 2000. How big are raindrops? *The Science Teacher* 67 (2): 46–50.

National Research Council (NRC). 1996. *National science education standards*. Washington, DC: National Academies Press.

Internet Resources

Bentley Snow Crystal Collection
http://bentley.sciencebuff.org/collection.htm
SnowCrystals.com
www.its.caltech.edu/~atomic/snowcrystals
The Snowflake Man
http://snowflakebentley.com/mullet.htm
Wilson Snowflake Bentley: Photographer of Snowflakes
www.snowflakebentley.com

This article first appeared in the Summer 2007 issue of Science and Children.

Chapter 22

Bubble Shapes

by Kathleen Damonte

Few can resist a bubble wand and a brightly colored bottle of bubble solution. Although playing with bubbles might seem like just a fun outdoor activity, it is also an opportunity to explore some interesting science concepts related to soap, light, and color.

What do you think is inside a soap bubble? If you guessed air, you are correct. Soap bubbles are air surrounded by a thin film of water and soap. The soap film wraps around the air and traps it within. The soap film is elastic—that means it is stretchy. Bubbles that are floating freely in the air are usually *spherical* (round). A sphere is the smallest surface area that can contain the air inside with the least amount of stretching for the soap film. A bubble that is blown

Further Reading

- "Bubble Makers," from *A Head Start on Science* (2007)
- "Lighten Up Your Lesson: Matter, Optics, and Bubbles," from the March 2006 issue of *Science Scope*
- "Bubbles," from *More Picture-Perfect Science Lessons* (2007)

on top of a wet surface uses that surface as a wall and will contract to form a dome shape.

BUBBLE COLORS

The colors you see in a soap bubble come from *white light*. White light is actually composed of seven colors: red, orange, yellow, green, blue, indigo, and violet. When white light passes through a transparent material, such as the soap film that makes up the bubble wall, the different colors separate so they can be seen. A bubble wall is extremely thin, just a few millionths of an inch thick. When a light wave hits the outside of the bubble wall, some of the light is *reflected* back to your eye. Light waves are also reflected back from the inside of the bubble wall. These reflected waves interact in a complicated process that produces the colors you see. As the bubble wall gets thinner and thinner, the bubble loses color. A bubble will look black at the top just before it pops.

WHY DO BUBBLES POP?

Bubbles pop for several reasons. Bubbles pop when the water in their bubble wall evaporates. When bubbles are blown outside in the sunlight, they evaporate more quickly. They also pop because of contact with wind, a dry surface, or dry air. To have the best results when working outdoors with bubbles, work in an open, shady area and keep all materials wet with bubble solution.

Experiment with bubble domes by trying the following activity.

Resources
Bubbles
www.exploratorium.edu/ronh/bubbles/bubbles.html

Bubble Fun
www.reachoutmichigan.org/funexperiments/quick/bubble-fun.html
The Art and Science of Bubbles
www.cleaning101.com/sdakids/bubbles

This article first appeared in the May 2003 issue of Science and Children.

Exploring Bubbles

Materials
- Cookie sheet with sides
- Bubble solution
- Plastic straw

*You will probably want to do these activities outside; it will make the cleanup a lot easier. Pick a shady area with little wind.

For this activity you can use prepared bubble solution from the store or mix up some of your own bubble solution from the recipe below.

Bubble Solution Recipe
- 0.95 L (1 qt) of water
- 120 ml (8 tbsp) of liquid dishwashing detergent (Joy works well)

Mix the two ingredients together in an appropriate-size container.

Directions
1. Fill the bottom of the cookie sheet with bubble solution so the entire bottom is wet.
2. Dip the bottom end of the straw in the bubble solution.
3. Touch the straw at a 45-degree angle to the bottom of the cookie sheet and blow gently to create a dome.
4. Be patient—it may take a few tries to learn how to blow a bubble dome.

Questions
1. Blow a bubble dome and touch it with a dry finger. Blow another bubble dome and touch it with a wet finger. What happens each time?
2. Blow the largest bubble dome you can. How were you successful in doing this?
3. Blow a bubble dome inside a bubble dome. How were you successful in doing this?
4. Blow two connecting domes, three connecting domes, etc. What happens where the bubbles join each other?
5. Blow a bubble dome and watch it carefully. What colors do you observe and how do they change? Try to predict when the bubble will pop.

Chapter 23

Helicopter Seeds and Hypotheses ... That's Funny!

by Leslie Wampler and Christopher Dobson

The most exciting phrase to hear in science, the one that heralds new discoveries, is not "Eureka! I found it!" but rather, "Hmmm, that's funny!"

—Isaac Asimov

Investigating maple samaras, or helicopters seeds, can give students a "that's funny" experience and catalyze the development of inquiry skills. In this article, we describe how to use maple helicopter seeds (samaras) to engage students in focused observation and hypothesis testing. This activity requires only basic classroom equipment and maple samaras, which can be found throughout most of the United States or purchased online.

FLYING FRUIT

Science comes alive for students when it coincides with play. Many students have spent time playing with maple helicopter seeds, also known as whirlybirds. Capture your students' interest by holding up one of these familiar seeds and asking, "Where do these come from? What happens when they fall?" If students are unfamiliar with samaras, give small groups several to observe. Even those who have played with helicopter seeds may be unaware that they contain the seeds of maple trees and are also known as *samaras*. Most, however, will be able to predict that the seeds will twirl or spin as they fall to the ground. At this point, toss the seed into the air and let the class observe its behavior. Build student confidence by celebrating their accurate predictions.

Ask students why they think samaras spin. They may hypothesize that the spinning helps them fly or ride the wind. Have students brainstorm how this motion might benefit maple trees. They should recognize that the helicopter's structure allows for wind-powered seed dispersal. Remind students that plants, like animals, have structures with specialized functions. Fruits are structures that flowering plants produce to disperse their seeds. Some fruits are edible and dispersed by animals. Samaras are dry fruits, specialized for wind dispersal. Maple trees have double samaras (Figure 1), each containing one seed, that usually fall separately.

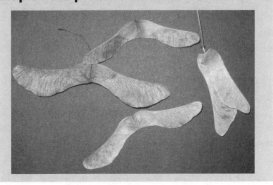

Figure 1

Maple trees produce double samaras.

Now the stage is set for further exploration. Students have drawn on previous knowledge and probably think they know how the samara works. So let's dig deeper.

DOES SIZE REALLY MATTER?

Maple samaras come in a variety of shapes and sizes (Figure 2). It takes more energy to make large samaras. Use guided discussion to encourage students to think about why a plant invests that extra energy and what advantage size gives to a seed. Remind students that seeds contain an embryo (or baby plant) and the food source needed for germination. Large seeds can carry bigger embryos and more nutrients. But what about the wing? Why do maple trees allocate additional energy to the samara's wing? Student responses often relate to the length of time samaras can stay in the air, giving them more opportunity to be carried by the wind.

Does size affect the length of time a samara remains airborne? Let's find out. Hold a "reverse race" between two single samaras of different sizes. The one that stays aloft longer wins. Why would this be an advantage in nature? Hold up the two samara racers. "Ready ... Set ... Wait!" Before dropping the seeds, remind students that science requires focused observation, or thinking about what we see. In math, we ask students to estimate to check their answers; in reading, we ask them to predict what will happen to build reading skills. In science, we ask students to observe carefully, to see with the brain engaged. How can we spot the unusual if we don't know what we expect to see? Scientists are good observers. Science also involves the formation of hypotheses, or possible explanations. Hypotheses are the basis for specific predictions, like whether the large or small samara will stay aloft longer when released simultaneously from the same height. The hypothesis supplies the rationale behind the prediction. Have students provide explanations (hypotheses) for their predictions (Figure 3). Most students predict that the small samara will stay aloft longer because it is lighter, or the big samara will stay in the air longer because of its larger wing. Make sure students understand that predictions allow us to test hypotheses. If the prediction is right, their hypothesis is supported; if not, their hypothesis is falsified.

Finally, we're ready to race. Hold each samara above your head at the same height, gripping the seed with the wing pointing down. Ask a few student volunteers to confirm simultaneous release (controlled variable). Release the seeds and watch as they twirl downward.

Figure 2

Samaras grow in a variety of shapes and sizes.

PHOTOGRAPHS COURTESY OF THE AUTHORS

TESTING HYPOTHESES

As a class, discuss these questions: Does this outcome confirm our hypothesis? Are there other possible explanations for the result? Is one trial sufficient? Suggest that students investigate for themselves by conducting their own races. Give small groups several large and small samaras for their exploration. Encourage students whose previous predictions were incorrect to show that the result was a one-time fluke and that their hypothesis really is correct. Inform students that wise scientists design experiments with the potential to falsify their hypothesis. If they are unable to falsify it, they tentatively accept the hypothesis, very aware that they may have missed the one experiment that would not support it. This is the tentative nature of science—hypotheses can be strongly supported, but at what point have they been confirmed?

One experiment with the potential to falsify the mass-related hypothesis—small samaras will stay aloft longer because they are lighter—involves separating the seed from the wing in both the small and large samaras. Have students race the two seeds without the wings, sharing their predictions first, as they did with the samara race. If the heavier seed reaches the ground first, the hypothesis is supported. If not, it must be rejected. Use scissors to separate the samaras' wings and seeds. To help students remember which seed and wing came from the larger samara, mark both pieces with a different color dot. Mass the seeds using a balance that can display milligrams. With wings removed, gravity is the primary force influencing downward motion of the seeds. Because acceleration due to gravity is not dependent on mass, the seeds should reach the ground simultaneously if they are released at the same time. Therefore, the mass-related hypothesis is not supported.

If students seem unimpressed, or if some seem to be clinging to the hypothesis that heavier objects fall faster, hold up a sheet of notebook paper and a thick textbook. Ask students which will hit the ground first and why. Crinkle the paper into a ball and drop it at the same time as the book. They will hit the ground at the same time. Clearly, acceleration due to gravity is not dependent on mass.

Because dropping seeds gave a different result than dropping samaras, what would happen if two different-size wings were raced? Students will find that the wing with the larger surface area stays aloft longer, even though it weighs more. Hmmm . . . that's funny! Now your students are ready to use the gray matter between their ears.

WHAT NOW?

At this point, few students will have a hypothesis that fits what we've observed. What will your students suggest the class try next? Encourage questions and suggestions. New discoveries are often the result of curiosity. What else can your students discover about how the parts behave differently from the whole?

We found that small samaras usually stay aloft longer than large ones, although there can be variation in the outcome, depending on release technique. The small samaras that we measured also happened to have lower wing loading (total weight/wing area), a characteristic that results in more lift in aircraft. In plants, rates of descent are correlated with the square root of wing loading for a variety of wind-dispersed species (Augspurger 1986). The upshot is that neither mass nor wing size alone accounts for the smaller samara's advantage in staying airborne.

Tell students that, when stumped, scientists often return to the observation phase. In other words, it's playtime again! Provide students with rulers and balances, and have them investigate the mass, surface area, and "flight patterns" of intact samaras, as well as their separated constituent parts (seeds and wings). As students apply their freshly honed powers of observation, they will find that samaras spin, wings float down, and seeds merely fall.

Compile class data on the board. Ask for interpretation of the data. Students will see that most of the intact samara's mass is in the seed, while the wing contributes the majority of surface area. The autorotation, or spinning, of the samara results from the combination of the wing's surface area and the off-center concentration of mass in the seed. When samaras fall from a low height without spinning, gravity is the primary force at work

---

Figure 3

Testing your hypothesis

Question: Will a large or small samara stay aloft longer?
Prediction:

Hypothesis (possible explanation for prediction):

Another way to look at this is if your hypothesis is true, then your prediction will occur. Reminder: If your prediction does occur, your hypothesis may still be incorrect—you haven't necessarily proved it!

---

and they reach the ground at nearly the same time. See Walker (1981) for a thorough explanation of maple samara aerodynamics.

Ask students for examples of how flowering plants produce fruits to disperse their offspring (seeds) and how different types of fruits are specialized for different types of dispersal (animal, wind, water). In the case of the maple samara, the structure of the fruit (wing) is marvelously designed to accomplish its function of flight. The longer the samara can stay airborne, the greater chance it has of being dispersed from the parent tree by the wind. This is a wonderful example of structure fitting function, one that your students are not likely to forget anytime soon!

EXTENSIONS

Have students manipulate helicopter mass and surface area by applying pieces of tape and clipping the wings, respectively, and then predict the impact on rates of descent (Thomson and Neal 1989).

Connect across the curriculum as highlighted in *Science Scope's* October 2007 history-of-science issue. Investigate lift and drag using the storyline approach of the Wright brothers and the history of flight (Isabelle 2007). Or use a literature-circle approach, incorporating biographies of the Wright brothers and their invention of flight (Straits 2007).

Acknowledgments
We would like to acknowledge Dayna Malcolm, Jessica Lilly, and Stephani Johnson for their help with the initial development of this lesson.

References
Augspurger, D. 1986. Morphology and dispersal potential of wind-dispersed diaspores of neotropical trees. *American Journal of Botany* 73 (3): 353–363.

Isabelle, A. D. 2007. Teaching science using stories: The storyline approach. *Science Scope* 31 (2): 16–25.

National Research Council (NRC). 1996. *National science education standards*. Washington, DC: National Academies Press.

Straits, W. 2007. A literature-circles approach to understanding science as a human endeavor. *Science Scope* 31 (2): 32–36.

Thomson, J. D., and P. R. Neal. 1989. How-to-do-it wind dispersal of trees seeds and fruit. *The American Biology Teacher* 51 (8): 482–486.

Walker, J. 1981. The amateur scientist: The aerodynamics of the samara: Winged seed of the maple, the ash and other trees. *Scientific American* 245 (4): 226–236.

This article first appeared in the September 2008 issue of Science Scope.

Chapter 24

The Egg

In the Lab and Across Cultures

by Judith J. Paolucci

Science teachers often assign essays about scientists from other cultures in order to address the issue of tolerance. Such assignments concentrate on the differences rather than the similarities among cultures. Often, understanding and tolerance are not achieved. Focusing on similarities among cultures brings about greater understanding and tolerance. An effective lesson I have used with students in a suburban middle school, as well as in inner city and suburban high schools, employs the egg as both the focus of a discussion about culture and an object in scientific demonstrations. I plan this lesson for the day before spring vacation, when students' thoughts are diverted toward time off and traditional family gatherings.

Besides focusing on the science behind a series of demonstrations involving eggs, I also address the cultural significance of eggs. At the start of the lesson, I share my research on eggs (see The Cultural Significance of Eggs, p. 112) with my students and they share their family traditions with me. If time permits, you can assign students a country to research and have them share their findings with the class through handouts or presentations.

Regardless of their religious or cultural backgrounds, students are unsure why eggs are associated with spring holidays. Most students are also surprised to find out that eggs are not exclusively part of a Christian tradition but are part of spring celebrations for many cultural and religious groups. As we talk about the historical and cultural importance of the egg, students add their own knowledge and traditions to the discussion. Students of different cultural backgrounds learn that while their families may have unique traditions, they also share some common beliefs and customs with the families of their peers.

EGG-CITING ACTIVITIES

This lesson, of course, would not be a science lesson without some science content! I chose these demonstrations based on the curriculum of the course I am teaching. (Note: These demonstrations can also be converted to student activities, except for those contraindicated by safety concerns.) Many of these experiments are classics—they have been passed from teacher to teacher and can be found in any number of books on demonstrations. A search on the internet will also reveal many sites that include egg demonstrations.

AN INCREDIBLE FOOD

Prior to the demonstrations, I share some the following egg facts with my students. The egg contains all that a baby chick needs to make feathers, bones, blood, a

The Cultural Significance of Eggs

The symbolic significance of an egg derives from the seemingly miraculous proliferation of life. Both eggs and bunnies (another symbol not exclusively Christian) are symbols of fertility and rebirth. People colored eggs in China, ancient Greece, and Rome for centuries before the birth of Christ. The egg is also associated with ancient religions and tribal customs. For example, in the ancient language of Egyptian hieroglyphics, the symbol of an egg indicates a life-giving seed or the mystery of life. Hindus consider eggs to be the source of life and do not eat them.

There are other spring holidays, besides Easter, in which eggs are part of the traditional feast. In Egypt, *Norooz* (the Persian New Year) is celebrated on the first day of spring. A ceremonial table called *Sofreh-e Haft Sinn* (cloth of seven dishes) includes dishes beginning with the Persian letter *Sinn*, along with other elements and symbols, including a basket of painted eggs representing fertility. For Jews around the world, Passover is a celebration of the Hebrews' escape from Egypt. The Passover Seder includes five foods that have special meaning for this holiday, including a roasted egg, symbolizing spring.

The early Christians adopted the egg as a symbol of the resurrection of Christ and it became part of the Easter celebration. There are different Christian traditions in different parts of the world. In Russia eggs are painted in different bright colors, although the red egg is considered a symbol of Easter. Russians also make eggs of different materials, like the porcelain Fabergé Imperial Eggs.

In France the custom of offering eggs as Easter gifts began in the fourth century AD. Church law dictated that Christians abstain from eating meat or eggs during the 40 days of fasting prior to Easter, known as Lent. Egg farmers would find themselves with an overabundance of eggs. On the days just prior to Easter, groups of children would form processions through the streets, collecting the surplus eggs donated by the farmers.

Bulgarian, Greek, and other orthodox Christians all over the world use the bright-red colored egg as the symbol of Easter. These hard-boiled eggs are cracked after the midnight service and over the next days. In one ritual each person selects his or her egg. Then people take turns tapping their egg against the eggs of others until one breaks. The person who ends up with the last unbroken egg is believed to have a year of good luck.

beak, skin, etc. (This usually brings out more than one cry of "Yuck!") The yolk is 49% water, 31.6% fat, 16.7% protein, 1.5% minerals, and vitamin A; while the albumen (white) is nearly 90% water, with protein, minerals, and riboflavin composing the rest. Due to the long strands of protein in the albumen, egg whites can be whipped to a foam, emulsified, or used as a gelling agent or to bind other foods together. (Earlier in the year, we study polymer chains, molecules having long strands, so students have a basic understanding of their form and function.) These are great properties for a baking and cooking ingredient.

DENATURING THE PROTEIN IN EGGS

The white of an egg (and the white of your eyes) is mostly protein and water. The structure of protein molecules changes, or denatures, when proteins react with acids. This demonstration involves cracking a raw egg into a beaker containing an acid and serves as a strong reminder of the importance of wearing goggles. When the albumen hits the vinegar, it will turn white as if it were being cooked. This reaction and potential damage to your eyes is irreversible. Safety goggles must be worn to protect the whites of your eyes during this demonstration. Do not allow students to perform this as an activity.

REACTION OF CALCIUM CARBONATE AND ACID

The eggshell is mostly calcium carbonate. When placed in vinegar, the acetic acid reacts with the calcium carbonate in the eggshell and dissolves the shell. All that's left holding the egg together is the membrane. It could take up to a week to dissolve the entire shell, but some vinegars may dissolve the shell overnight. This experiment can also be done at home, but students will need to take care when transporting the shell-free eggs back to the classroom for the osmosis activity that follows.

When eggshells react with vinegar, a gas is formed. You can verify that the gas is carbon dioxide through a limewater test. Place bits of eggshell in an Erlenmeyer flask and add vinegar. Top the flask with a one-hole stopper connected to a piece of rubber tubing. Place the other end of the tubing in a beaker of limewater. As the gas bubbles through the limewater, the liquid will turn cloudy, indicating the presence of carbon dioxide. Alternatively, you can direct the flow of gas at a lit match or candle to extinguish it.

OSMOSIS

The egg's membrane, which coats the inside of the shell, is similar to a cell membrane because it can selectively allow substances in or out. Osmosis is responsible for the movement of small nutrient molecules and water into cells, for the movement of waste products (such as urea) out of the cell, for the production of pickles from cucumbers, and so forth. The pressure that counteracts the movement of water across such a semipermeable membrane is called *osmotic pressure*. In this demonstration students use the "naked" eggs that were created through the reaction of the eggshell with vinegar. One naked egg is placed in a beaker containing a diluted clear corn syrup solution (50:50) two days prior to the day of the demonstrations. A second egg is placed in a beaker containing saltwater. Additional eggs are added to solutions of food coloring. At the end of the two days, students observe the eggs to see what was absorbed or expelled. The diameter of the eggs and other physical characteristics should also be recorded.

BALANCE

Why did Humpty-Dumpty fall down? Eggs cannot be balanced too well because the inside of an egg is a very viscous (thick) liquid, and the yolk sits in this liquid. The yolk is usually a bit off center, and rides high in the egg, making it very difficult to balance. This makes the egg fall over. There is a superstition that it is possible to stand an egg upright on its end on the date of the equinoxes (and/or solstices) because universal forces are in balance on these days. This is nonsense! There is nothing special about the conditions on these days. You actually can balance an egg on its end at any time of the year if the egg has some little bumps on its end and if you are very patient.

Students love trying to balance eggs. I show them that I can do this easily, though I don't tell them right away that I have prepared my "obedient" egg by blowing out its contents and filling the egg a quarter full with fine sand. Using wax, I seal the egg openings so that it appears to be a normal egg. Because the sand is free to move in any direction inside the egg, this permits the egg to balance in whatever position you place it regardless of the angle.

In an additional demonstration, we investigate the difference between the spins of a raw egg and of a boiled egg. First I spin a hard-boiled egg on its side, and then bring it to a complete stop by pressing down on it with one of my fingertips. Next, I spin a raw egg on its side, quickly make it come to a dead stop by pressing on it with one of my fingertips, then immediately remove my finger. Students are surprised to see the egg begin to spin slowly thanks to the inertia of its rotating interior. Before letting students try this on their own, have them try to explain why the two eggs behaved differently.

DENSITY

Eggs are more dense than freshwater, but less dense than saltwater. I pretend to teach eggs to read by writing *float*, *sink*, and *swim* on three identical eggs. I then place these eggs in 500 ml graduated cylinders containing saltwater, freshwater, or layers of saltwater and freshwater. The egg floats in saltwater, sinks in freshwater, and swims between the layers of freshwater and saltwater.

THE STRENGTH OF EGGS

I place four eggs of approximately equal size in the second and fifth rows of an open egg carton, with their pointed ends down. After closing the carton, I carefully place textbooks directly over the eggs, one at a time. I have stacked 20 textbooks in this way without breaking the eggs. The egg's shape—two domes pushed together—makes it resistant to compression but weak in tension (see above). Students discuss why this property is useful. Students may suggest that eggs' inherent strength protects their embryo, while its inherent weakness allows the chick to peck its way out.

ACIDS AND BASES

Prior to the lesson, I paint designs on white eggs with phenolphthalein, a common acid-base indicator. During the lesson I spray these eggs with a diluted solution (approximately 0.1 M) of sodium hydroxide. The sodium hydroxide causes the phenolphthalein to turn a bright pink, indicating the presence of a base. Because of the nature of the chemicals involved, this should only be performed as a demonstration by a teacher wearing safety goggles.

This article first appeared in the April 2003 issue of Science Scope.

Chapter 25

An Outdoor Learning Center

by the Natural Resources Conservation Services, USDA, and NSTA

LESSON DESCRIPTION

With adults' help, students inventory the school site, develop plans, then create a garden. This raises awareness among students, teachers, and parents about the natural environment and about using the school site for hands-on learning.

Subjects

Art, language arts, mathematics, science, social studies

Time

Prep: 2 hours minimum
Activities: 4 ½–10 hours (not including Extensions)

Further Reading

- "Building Bird Nests," from A *Head Start on Science* (2007)
- "Nature Bracelets," from A *Head Start on Science* (2007)

TEACHER BACKGROUND

This lesson provides students and teachers with an ongoing opportunity for hands-on environmental education and *resource conservation.*

A readily accessible resource for teaching is the school site. An outdoor learning center (OLC) on the school site offers educators and students an exciting place to observe nature's happenings through the seasons. Right outside the classroom, the school site offers many opportunities to publicize conservation in the neighborhood by improving students' knowledge about and concern for the natural world. By implementing an OLC on the school site, the teacher can maximize teachable moments relating to the environment and *natural resources.*

If your school or neighborhood already has an OLC, skip ahead to the "OLC Activities" section below.

Enlisting Assistance

Begin the planning process as early as possible. It is important to secure the school administration's permission and obtain support from the school's maintenance staff for your project. If you teach young children, find a teacher of older grades who shares your interest in the project. This teacher's students and yours can become teammates or buddies for the school project. Remember to start small: The project has a greater chance of succeeding if original goals are modest and leave opportunity for growth. As members of the school community see the success of this first step, they may provide support for an OLC for the entire school.

If your school site project has the space and has been well planned, it may be easily adapted for additional outdoor learning activities in continuing school years. If this idea is approved by the administration, you may tell students, parents, and other teachers that this project will be the first step in establishing an OLC on the school site—a place to do hands-on activities, learn about the environment, and participate in actual resource conservation projects. Remember that you are dealing with natural as well as human influences. Be prepared to explain limitations of the OLC, such as temperature, moisture, insects, wind, limited space, or the wrong soil. Learning from this year's activities can help create a more successful OLC next year.

There are many ways to solicit the equipment needed to create and maintain your OLC. Team up with a high-school agricultural program and share supplies. Apply for a grant with a local or national gardening or environmental education association. Ask a landscape firm, local business, or government agency to donate tools.

Creating an OLC Garden

A common and effective school site activity is establishing a garden, which is an environment that students can manipulate. Students' planning, planting, and caring lead to the excitement of harvesting the rewards of their efforts. You can establish a vegetable or flower garden almost anywhere: in a large or small space, on a flat area or on a slope, in the shade or in full sunlight, on the school roof, or on a narrow strip of land between a parking area and the school building. Whichever area is used, the OLC garden provides a venue for short- and long-term environmental learning.

To create a garden, first analyze the site. Students should observe and record the site's physical and environmental characteristics. This class survey provides a starting point and helps show the changes that take place over time. Document modifications to the OLC garden to provide a compete record. After the site has been analyzed, discuss planting, maintaining, and harvesting a garden. The class can then decide on the type of flowers or vegetables to grow, design the garden layout, and plant.

Materials

Actual materials required for this activity depend on the needs identified through the inventory and planning. A soil survey determines the soil type of the school site and helps you select the correct vegetation. Use native plant species whenever possible, since they tend to require less water, weeding, and fertilizer than exotic species. Be sure that none of the plants are invasive, especially if your site is near any natural areas. You can obtain this information from local soil and water conservation agencies.

This activity offers a valuable opportunity to stress safety with your students. Emphasize the correct way to use and treat tools. Ask students to wear pants and shirts with long sleeves on days when they are working outside to avoid insect bites and irritating plants. Teach students what poisonous plants look like and how to avoid such plants. Find out in advance which students have insect and plant allergies, and take necessary precautions.

OLC Activities

The activity, observation, and records of an OLC should be continual and should demonstrate interrelationships between humans and the rest of the natural world. Activities should be inquiry-based and lead to the resolution of issues.

The following suggestions for OLC activities focus on *conservation, beautification,* and wildlife *habitat* improvement:

- Adopt a section of the OLC, a playground, or a nearby stream. Remove all trash and keep the area clean.
- Plant trees or shrubs that shelter the school site from the wind.
- Plant flowers, trees, or grasses to stop soil erosion.
- Invite birds to your area by adding birdhouses near shrubs or trees that provide protection from predators and by choosing plants that provide food and shelter (see Figure 1).
- Create a butterfly garden by using plants and flowers that attract butterflies.
- Order vegetation native to your area and plant a natural landscape.
- Plant grass and trees that are valuable for shade, nesting, and beauty and that vary in color, texture, and shape.
- Adopt a special tree and note seasonal changes, animals that live in the tree, and outstanding characteristics of the tree using photos, drawings, and writing.
- Identify rocks or boulders on the site. Investigate the types of materials used to build the school and compare materials to the rocks on the site.

- Examine a rotting log to observe fungi, moss, and insects.
- Record temperature, wind, or precipitation over time, and then graph the data.

You might begin by having students classify the environmental events that take place on the school site on a regular and seasonal basis. Students can pass records to succeeding classes to build an environmental history of the site. Over time, students might chart differences in rainfall, snowfall, temperature, growth and death of plants, or erosion. Older students could research the history of the school site. By analyzing history and environmental events through tables, graphs, and written logs, students become more aware of the school site environment.

LEARNING CYCLE

Student Objectives
Students will be able to

- design and build an OLC garden,
- justify the importance of their school-site conservation activities, and
- explain some of the activities or events in the OLC garden.

Materials
For the Class
- Poster board
- Marker
- Local soil survey
- Plastic transparencies
- Overhead markers
- Gardening tools (e.g., hoes, rakes, spades)
- Work gloves
- Plants, trees, and shrubs
- Hose
- Camera (optional)

For Each Student Group
- Diagram of the school site
- Pencil
- Writing paper
- Drawing paper

Figure 1

Plants that provide food for wildlife

Trees	Shrubs	Flowers
Oak	Viburnum	Cosmos
Black Walnut	Blueberry	Impatiens
Crabapple	Dogwood	Marigold
Maple	Lilac	Zinnia
Pine	Sumac	Phlox
Spruce	Pfitzer Juniper	Trumpetvine
Desert Willow	Ocotillo	Desert Baileya
California Buckeye	Desert Hackberry	California Poppy

Some of these plants may not be appropriate for your region. Avoid using nonnative plant species.

Perception: 30 minutes–1 hour
1. Introduce students to the idea of an OLC.
2. Begin planning the project by brainstorming ideas for a garden. Let students lead by providing ideas and making notes on the board. What are the students' desires and concerns for the garden? Encourage students to discuss their ideas about planning and placement, and illustrate those ideas on the board and record them.

Exploration: 30 minutes–2 hours
Sketch a simple diagram of the school site.

1. Take the class outside to map the school site. Distribute diagrams of the school site and have students record the physical characteristics of the site. For instance, you might ask students to map areas of bare soil, direct sunlight, vegetation, pavement, and buildings, and compare the slope of the ground in various places. To save time, you can assign each student group to map one characteristic of the site. Then transfer all the maps to clear transparencies and overlay the maps for an overview of the school site.
2. After students have created their maps, suggest to students how you could use this information to create a successful garden. Discuss how students' project ideas would work with the school site's available space. Adjust the plan, as ideas are accepted. This organizing session allows students to communicate, plan, and be responsible for the development of their own school site project.
3. To actively involve older students in the planning process, hold a contest to select the best plans

for the garden. Divide students into groups of two or three and ask groups to draw up plans and materials lists of their ideas about what the garden should look like. A panel of teachers, administrators, maintenance staff, and older students choose the top three plans. The class then votes for its favorite plan of the top three.

4. Using the class's suggestions, draw a plan of the garden on poster board.

5. When the project has been finalized, type or print all relevant information and create the formal plan for the school site project.

Application: 3–6 hours

Planning, organizing materials, getting permissions, and involving parents, students, and school staff takes several hours. Be sure the adults don't take over the project; this should be a fun and exciting time for student discovery. Also, remember that there is no deadline—this project may never be finished. Ideally, the excitement generated by the school site project will encourage duplicate efforts in the school community and the community at large, starting with home gardens or other beautification activities.

1. Take the class outside and demonstrate the proper use of gardening tools.

2. Split students into small groups or pair students with older teammates or buddies.

3. Assign group roles and responsibilities. Some teams can begin planting while other groups sketch or list more ideas for the garden.

4. Every student should have the opportunity to do some type of gardening activity—raking, planting, etc. Such active participation gives students a sense of ownership for the program and helps them develop a sense of belonging and personal satisfaction.

Evaluation: 30 minutes–1 hour

Evaluation should be an ongoing process as the school site project is developed and includes formal follow-up with students, parents, and other school staff. The

students should be allowed to express suggestions for the next phase of the school site project.

Extensions: 30 minutes each, minimum

- Read a story about planting a garden.
- As a class, discuss ideas for expanding the current school site project, develop a plan, and present it to the school administration.
- If an OLC is not possible at your school, identify and label the vegetation currently growing on your school campus. Students can also observe wildlife on your school grounds or in a nearby park.
- Students can grow vegetables or flowers in a garden, using stakes to identify each plant. Choose plants that will produce results before the school year ends.
- If you grow produce in your school garden, try these ideas:

  - Invite a parent to prepare some of the produce grown in the garden.
  - Give produce to the school cafeteria to use in a meal for students.
  - Donate produce to a homeless shelter or soup kitchen.
  - Allow students to divide and take home any produce or flowers.

- Each student can pick a plant in the garden and measure and graph its growth over time. Students can also draw the plant in various stages of growth, or through the seasons.
- Establish a nature trail on or near your school site.
- Take a field trip to a farm or garden center to see how "big" gardens are planted and cared for, or invite a farmer or garden center employee to class to share expertise, experience, and perhaps some plant materials or tools with students.

This chapter first appeared in Dig In! Hands-On Soil Investigations *(2001), from NSTA Press and the Natural Resources Conservation Service, USDA.*

Chapter 26

Be a Friend to Trees

by Karen Ansberry and Emily Morgan

DESCRIPTION

Learners explore the variety of products made from trees; the importance of trees as sources of food, shelter, and oxygen for people and animals; and ways to conserve trees.

SUGGESTED GRADE LEVELS: K–4

Lesson Objectives Connecting to the Standards

Content Standard A: Science as Inquiry: Abilities Necessary to Do Scientific Inquiry
- Ask a question about objects, organisms, and events in the environment.

Further Reading

- "Adopting a Tree," from *A Head Start on Science* (2007)

Content Standard B: Life Science: Organisms and Their Environments
- Understand that all animals depend on plants. Some animals eat plants for food. Other animals eat animals that eat the plants.

Content Standard F: Science in Personal and Social Perspectives: Types of Resources
- Understand that the supply of many resources is limited. Resources can be extended through recycling and decreased use.

TIME NEEDED

This lesson will take several class periods. Suggested scheduling is as follows:

Day 1: **Engage** with *Our Tree Named Steve* read aloud.
Day 2: **Explore/Explain** with Sorting Chart and *Be a Friend to Trees* read aloud.
Day 3: **Elaborate** with My Favorite Tree.
Day 4: **Evaluate** with *Be a Friend to Trees* Poster.

MATERIALS PER GROUP OF 3–5 STUDENTS

- Sorting chart made from chart paper with a large Venn diagram drawn on it

- Boxes or bins, one per group, filled with several of the following tree parts or products (actual objects or pictures of the objects) described in the book *Be a Friend to Trees*:

Featured Picture Books

Title: *Our Tree Named Steve*
Author: Alan Zweibel
Illustrator: David Catrow
Publisher: G. P. Putnam's Sons
Year: 2005
Genre: Story
Summary: In a letter to his children that is both humorous and poignant, a father recounts memories of the role that Steve, the tree in their front yard, has played in their lives.

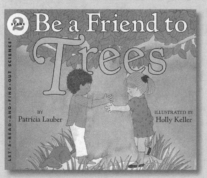

Title: *Be a Friend to Trees*
Author: Patricia Lauber
Illustrator: Holly Keller
Publisher: HarperTrophy
Year: 1994
Genre: Non-narrative Information
Summary: Discusses the importance of trees as sources of food, oxygen, and other essential things and gives helpful tips for conserving this important natural resource.

From Trees
- Wooden block
- Writing or construction paper
- Newspaper
- Small cardboard box or paper milk carton
- Apple, orange, pear, cherry, or peach
- Walnut, almond, pecan, or hazelnut in the shell (Check to see if you have students with tree nut allergies in your class, and use only pictures if you do.)
- Small tree branch with leaves
- Pine needles
- Piece of tree bark
- Paper towel
- Paper grocery bag
- Sealed baggie or balloon blown up with air and marked "Oxygen" (This will represent oxygen, although it also contains other gases.)
- Also include some of the following objects (actual objects or pictures):

Not From Trees
- Plastic objects (such as small toys, markers, balls, and containers)
- Metal objects (such as keys, foil, and spoons)
- Glass marble
- Rock
- Small pumpkin, squash, carrot, or potato
- Cotton, polyester, or nylon cloth
- Reusable net or canvas grocery bag
- Sealed baggie blown up with air and marked "Carbon Dioxide" (This will represent carbon dioxide although it contains other gases.)

Both
- Pencil with eraser
- Plastic bottle of maple syrup
- Chocolate in a foil wrapper

MATERIALS PER STUDENT FOR MY FAVORITE TREE ACTIVITY

- Crayon with the paper removed
- Pencil
- Clipboard (or notebook to use as a writing surface)

STUDENT PAGES

- My Favorite Tree journal (4 single-sided pages stapled together)
- *Be a Friend to Trees* Poster rubric

BACKGROUND FOR TEACHERS

Trees are one of Earth's most important natural resources. We depend on trees for food and wood products, water and soil conservation, shade, beauty, and, most important, the oxygen they add to the air. It is essential for students to understand and appreciate the importance of trees to humans and all life on Earth, and to realize that their actions can have an impact on trees. The National Science Education Standards state that children should have a variety of experiences to help them understand how animals, including humans, depend on plants and that the supply of natural resources like trees can be extended through decreased use and recycling. The Explore/Explain phase of this lesson allows students to explore our dependence on trees by observing and sorting various products that come from them. After reading about how humans and other animals depend on trees, they also learn a variety of ways that they can be a friend to trees.

Nurturing a sense of wonder about trees will encourage students to do more to protect and conserve this vital resource. The Standards suggest that children in the elementary grades build understandings of biological concepts through asking questions and through direct experience with living things, their life cycles, and their habitats. The Elaborate phase of this lesson involves students observing trees closely; noticing details in their shapes, leaves, and bark; and generating wonderings about their favorite trees.

ENGAGE

Our Tree Named Steve Read Aloud
Stop and Jot, Turn and Talk

Before reading the book *Our Tree Named Steve*, engage students by saying, "From where you are sitting, look around and think of everything in this room that might be different if there was no such thing as a tree." Allow some quiet thinking time, and then have students turn and talk to a neighbor. You may want to have students stop and jot their ideas before sharing with a neighbor.

Inferring

Explain that you have a book to share about a very special tree. Show the cover of *Our Tree Named Steve*, and then introduce the author and illustrator. Ask

•   What are you thinking this story is about? Why do you think so?

Synthesizing

Read the book aloud, stopping after page 5 (" . . . Mom and I got the hint and asked the builder to please save Steve.") Then ask

•   Now what are you thinking this story is about?

You may want to stop at key points in the story to allow students to discuss their thinking about the story's meaning.

Questioning

After reading, draw a large tree trunk on the board or chart paper and label it "Steve." Ask

•   How did the tree get its name? (The youngest daughter couldn't pronounce "tree" and called it "Steve.")

You can write the students' responses to the following two questions as "branches" of the tree. Ask

•   How did the family use this special tree when it was alive? (Answers can include as a swing holder, target, third base, hiding place, jump-rope turner, clothesline, hammock-holder, and sewer-water remover.)
•   How did the family use the tree after it blew over in the storm? (They used the wood to build a tree house.)

Then ask

•   How did the tree protect the family "to the very end"? (It didn't fall on their house, the swings, the dog's house, or the garden.)

Synthesizing
•   Now what are you thinking the story is about?
•   How does the story make you feel?
•   Have you ever had a special tree? What made it special?
•   What are some ways that trees help us?

EXPLORE/EXPLAIN

Sorting Chart and *Be a Friend to Trees* Read Aloud

In advance, create boxes or bins filled with an assortment of items that came from trees, items that did not come from trees, and items that contain both wood products and other materials (see materials list). Explain that students will be learning about some of the ways that trees help us by doing a sorting activity and then reading a nonfiction book. Divide students into groups, and distribute to each group a bin and a Sorting Chart made from chart paper with a Venn diagram (two large intersecting circles) drawn on it. Have students label one circle "From Trees," the other circle "Not From Trees," and the intersection of the circles "Both."

Then have groups observe each object carefully, discuss whether or not they think it came from trees and why, and place it in the appropriate circle on the chart. If they are not sure about how an object should be grouped, they can leave it in the bin for now.

Invite students to justify how they sorted the objects. Ask

- What are some of the objects you think came from trees? Why do you think so?
- What are some of the objects you think did not come from trees? Why do you think so?
- Were there any objects you were unsure about? Why?

Inferring

Next, show students the cover of the book *Be a Friend to Trees*. Ask

- What do you think this book might be about? Why do you think so?

Determining Importance

Tell students that *Be a Friend to Trees* is a nonfiction book that might help them learn which of the objects came from trees. Introduce the author and illustrator of the book, and then explain that as you read, you want them to listen for any of the items they placed in the "From Trees" circle on their sorting charts. Ask them to signal (raise hand, touch nose, or in some other way) when they hear about one of the objects.

Questioning

As you read aloud, stop periodically to question students to check for understanding and build interest. Some suggested questions are

- (p. 10) Look at the diagram. What is the first thing that happens in order to make paper? (Wood chips are cooked with chemicals.)
- (p. 10) What are the wood chips called after they become soggy? (pulp)
- (p. 10) What must be done to the pulp after the water is drained off? (It is dried, flattened, and then rolled into paper.)
- (p. 12) What are the only living things that can make their own food? (green plants)

- (pp. 14–20) What are some of the ways that animals use trees? Turn to a neighbor, and share an example from the book. (Possible answers include many animals eat leaves, bark, buds, and twigs; squirrels and chipmunks gather nuts to eat; bees collect pollen and nectar; birds roost and nest in trees; and deer hide beneath trees.)
- (p. 21) How do trees help the soil? (They keep it from washing away.)
- (p. 22) What would happen to people and animals if there were no trees or green plants? (There would be no oxygen in the air, and we couldn't breathe.)
- (p. 23) Where do trees make food? (in their leaves)
- (p. 24) What three things do leaves need to make food? (water, carbon dioxide, and sunlight)
- (pp. 30–32) What are some things you can do to be a friend to trees? Turn to a neighbor, and share an example from the book. (Answers might include use less paper; reuse paper bags; write on both sides of paper; recycle newspaper; and plant a tree.)

After reading, give students the opportunity to return to their sorting charts and move any of the objects to a different spot on the chart if necessary. Then discuss what kinds of things come from trees (such as oxygen, fruits, nuts, and wood and paper products) and what kinds of things don't (such as carbon dioxide, vegetables, plastics, metals, cloth, glass, and rocks).

ELABORATE

My Favorite Tree

This activity can be done on school grounds, during a field trip to a park or other wooded area, or as a take-home assignment. Take students outside to look closely at a tree. They will each need a copy of the My Favorite Tree journal, a clipboard, a pencil, and a crayon with the paper removed. First, model how to sketch a tree's shape and make careful observations of its leaves and bark.

Then, show students how to do a leaf rubbing:

1. Find a fallen leaf that is still soft, and place it on your clipboard with the rough or vein side up.
2. Place the journal page over the leaf.
3. Gently rub the long side of the crayon over the leaf.

Next, demonstrate how to do a bark rubbing:

1. Pick the part of the bark that you want to make a rubbing of.
2. Place the journal page over that part.
3. Gently rub the long side of the crayon over the bark.

Next, model some of your own wonderings about the tree. (For example: How old is this tree? I wonder who planted it. I wonder if an animal lives in this hollow part. What kind of tree is it?)

Finally, share your thoughts and feelings about the tree by explaining why you chose the tree for your journal. (For example: This is my favorite tree because the bark peels up in places and looks like paper. I like how I can fit my arms all the way around the trunk. I have never seen a tree like it before. I feel peaceful when I sit with my back leaning against the trunk.)

If this activity is to be done at home, students can take their journals home and complete them with an adult helper. If this activity is to be done at school or on a field trip, allow students to look at several trees before deciding on a favorite to include in their journals. When you return to the classroom, have students share their journals with one another.

EVALUATE

Be a Friend to Trees Poster
Ask

- What does it mean to "be a friend to trees"? (to do things that will help protect or conserve trees)
- Why is it important to "be a friend to trees"? (Answers could include trees help animals, humans, and the environment in many ways.)

Pass out the *Be a Friend to Trees* Poster grading rubric. Have students create a 3-2-1 poster summarizing what they have learned about trees and their conservation. Posters should include

- thorough descriptions of three ways trees are helpful to humans, animals, and the environment;
- two interesting facts about trees; and
- one labeled drawing showing a child being a friend to trees.

For fun and extra credit, students can include their own additional research on trees, or a poem, song, rap, or cheer about being a friend to trees. You can use the rubric to score completed posters and make comments.

More Books to Read

Gackenbach, D. 1992. *Mighty tree.* New York: Voyager Books.
  Summary: Three seeds grow into three beautiful trees, each of which serves a different function in nature and for people.

Gibbons, G. 1984. *The seasons of Arnold's apple tree.* New York: Voyager Books.
  Summary: As the seasons pass, Arnold enjoys a variety of activities as a result of his apple tree. Includes a recipe for apple pie and a description of how an apple cider press works.

Mora, P. 1994. *Pablo's tree.* New York: Simon & Schuster Books for Young Readers.
  Summary: Every year, Pablo's grandfather decorates a special tree for his birthday.

Shetterly, S. H. 1999. *Shelterwood.* Gardiner, ME: Tilbury House.
  Summary: While staying with her grandfather who is a logger, Sophie learns about different kinds of trees, what they need to thrive and grow, and how the bigger trees provide shelter for the smaller ones. Her grandfather teaches her that when harvesting trees, it is important to let the tallest ones stay to drop their seeds and start a new generation. Sophie discovers that when we take care of the woods, it provides for us for generations to come.

Silverstein, S. 1997. *The giving tree.* New York: Scholastic.
  Summary: Shel Silverstein's poignant story of a boy and a special tree that gives him many things throughout his life.

Udry, J. 1956. *A tree is nice.* New York: HarperCollins.
  Summary: This Caldecott award–winning book speaks simply and elegantly of the many pleasures a tree provides.

Worth, B. 2006. *I can name 50 trees today! All about trees.* New York: Random House.
  Summary: While stopping to admire some of the world's most amazing trees, the Cat in the Hat and friends teach beginning readers how to identify tree species from the shape of their crowns, leaves, lobes, seeds, bark, and fruit. Dr. Seuss–inspired cartoons and verses teach readers about many of the trees common to North America.

Websites
Arbor Day Foundation
  www.arborday.org
Trees for Life
  www.treesforlife.org

This chapter first appeared in More Picture-Perfect Science Lessons: Using Children's Books to Guide Inquiry, K–4 *(2007), by Karen Ansberry and Emily Morgan.*

My Favorite Tree

By _____

The Shape of My Favorite Tree

NATIONAL SCIENCE TEACHERS ASSOCIATION

My Favorite Tree cont.

Leaf Rubbing From My Favorite Tree

Observations of the leaf: _____

_____

_____

_____

_____

_____

My Favorite Tree cont.

Bark Rubbing From My Favorite Tree

Observations of the bark:_____

_____

_____

_____

_____

_____

NATIONAL SCIENCE TEACHERS ASSOCIATION

My Favorite Tree cont.

Wonderings about my favorite tree: _____

_____

_____

_____

_____

_____

_____

_____

_____

Why this is my favorite tree: _____

_____

_____

_____

_____

_____

_____

_____

_____

_____

_____

_____

Be a Friend to Trees

3-2-1 Poster Rubric

*Name*_____

Your poster includes:

3 ways trees are helpful to humans, animals, and the environment.

        1   2   3

2 interesting facts you learned about trees.

        1   2

1 labeled drawing of yourself "being a friend to trees."

        1

For fun and extra credit, you included your own additional research on trees, or a poem, song, rap, or cheer about being a friend to trees.

Total Points_____/6

Comments: _____

Chapter 27

Spiderweb Collecting

edited by William C. Ritz

INVESTIGATION

Observing a teacher collect an old or abandoned spiderweb

PROCESS SKILLS

Observing, comparing, communicating

MATERIALS

Hand lenses, spray paint, paper in a contrasting color to the paint, sticks (skewers) or rulers, spray misters (adjusted to produce a fine mist) (CAUTION: Use spray paint [latex] that contains low or no VOCs.)

PROCEDURE

Getting Started
Go on a walk to look for spiderwebs. Caution children not to touch the spiders. When you find one, check to see if the spider is nearby. If so, just observe the movement of the spider for a while. If not, spray the web with the spray mister to make it easier to observe. Then allow the children to look at a strand of the web with a hand lens.

Questions and Comments to Guide Children
- What is the shape of the spiderweb?
- It looks like there are bugs caught in the spider's web.
- Is a spider in the web? What is it doing? How do you think the spider made its web? What makes you think so?
- What is holding up the web?
- If you can, touch the web. How does it feel?
- Where does the spider hide if it is not in the web?

How to Preserve and Collect the Spiderweb
- Make sure the spider has left the web and is not around.
- Spray the front and back of the web with spray paint that contains low or no VOCs.
- Before the paint dries, place a piece of construction paper behind the web and carefully move the paper forward, making the web stick to the paper.
- You may want to spray the web and construction paper with a clear acrylic paint after it has dried to preserve it.
- Try to collect several different types of webs; hang them on the wall and compare them.

Follow-Up Activities

(1) Read a book about spiders to the children (one that shows pictures of spiders making a web). (2) Give each pair of children a small ball of yarn and allow them to make a spiderweb between the legs of a small chair turned upside down.

CENTER CONNECTION

Place several books about spiders in the center with yarn and a few branches that could be used to make spiderwebs in them.

LITERATURE CONNECTION

The Very Busy Spider
by Eric Carle, Putnam, 1984.

I Love Spiders
by John Parker, Scholastic, 1988.

Zoe's Webs
by Thomas West, Scholastic, 1989.

Spiders and Their Webs
by Darlyne A. Murawski, National Geographic Children's Books, 2004.

ASSESSMENT OUTCOMES AND POSSIBLE INDICATORS*

- **Science: Scientific Skills and Methods**
  Begins to use senses and a variety of tools and simple measuring devices to gather information, investigate materials, and observe processes and relationships.

- **Science: Scientific Knowledge**
  Expands knowledge of and abilities to observe, describe, and discuss the natural world, materials, living things, and natural processes.

- **Literacy: Book Knowledge and Appreciation**
  Demonstrates progress in abilities to retell and dictate stories from books and experiences; to act out stories in dramatic play; and to predict what will happen next in a story.

- **Creative Arts: Art**
  Progresses in abilities to create drawings, paintings, models, and other art creations that are more detailed, creative, or realistic.

- **Physical Health and Development: Fine Motor Skills**
  Grows in hand-eye coordination in building with blocks, putting together puzzles, reproducing shapes and patterns, stringing beads, and using scissors.

- **Physical Health and Development: Health Status and Practices**
  Builds awareness and ability to follow basic health and safety rules such as fire safety, traffic and pedestrian safety, and responding appropriately to potentially harmful objects, substances, and activities.

*Source: *Head Start Bulletin*. 2003. Issue No. 76. U.S. Department of Health and Human Services, Head Start Bureau. *www.headstartinfo.org/publications/hsbulletin76/hsb76_09.htm*

What to Look for

Not Yet	Child is not interested in observing a spiderweb.
Emerging	Child shows some interest; observes, asks one or two questions, and listens while others discuss observations.
Almost Mastered	Child observes spiderwebs, asks questions, and discusses observations.
Fully Mastered	Child makes observations, independently points out similarities and differences of spiderwebs, and asks several questions.

chapter 27

SPIDERWEB COLLECTING
Family Science Connection

Looking for spiders and their webs can be fun and fascinating but please **do not touch the spiders** you happen to observe. Go on a spiderweb hunt and see if you can find the **different sizes and shapes** that make up webs. As you and an older member of the family find webs, **notice** what they are connected to, if bugs are trapped in the web, and if the spider is weaving more web. **Compare** the spider's size to the web size. (CAUTION: In your hunt, avoid going into old buildings or heavy foliage sites—for example, old sheds or big, deep bushes—where poisonous insects and poisonous plants might be found. Instead, check a simple garden with flowering plants, or even fence posts.)

Comments or questions that may add a *sense of wonder* to this activity:

- Do bigger spiders make bigger webs?
- What is the spider doing?

COLECCIONANDO TELARAÑAS
Ciencia en familia

Buscar arañas y sus telarañas puede resultar divertido y fascinante, pero por favor **no toquen las arañas** que observen. Vayan en busca de telarañas y vean si pueden encontrar las **diferentes medidas y formas** en que están hechas las telarañas. Cuando ustedes y un miembro mayor de la familia encuentren telarañas, **fíjense** a qué está conectada la telaraña, y si la araña tiene insectos atrapados en la telaraña, o si la araña está tejiendo más en la telaraña. **Comparen** la medida de la araña con el tamaño de la telaraña. (PRECAUCIÓN: En su búsqueda, evite ir a edificios viejos o sitios con plantas tupidas—por ejemplo, depósitos viejos o arbustos grandes y profundos donde insectos y plantas venenosas pueden ser encontrados. En vez, busque en un jardín simple con plantas en flor o aún en verjas.)

Comentarios o preguntas que pueden *despertar curiosidad* en esta actividad:

- ¿Tienen las arañas más grandes telarañas más grandes?
- ¿Qué está haciendo la araña?

This chapter first appeared in A Head Start on Science: Encouraging a Sense of Wonder *(2007), edited by William C. Ritz.*

Chapter 28
A Garden of Learning

By Tasha Kirby

In the fall of 2005, a fellow teacher commented how the space in front of our portable classrooms was not only barren but also an eyesore. After brainstorming creative ways to use the area, we decided that a native plant garden would be the best way to beautify the area while furthering student learning. Our school already had two established gardens (vegetable/flower and butterfly) and a wetland, so our garden would expand on those learning resources while providing a tie-in to social studies. In creating the native plant learning garden, our fourth-grade students were responsible for designing a layout, researching garden elements, preparing the area, and planting a variety of native plants.

Further Reading

- "An Invertebrate Garden," from the Summer 2008 issue of *Science and Children*
- "Wiggling Worms," from *More Picture-Perfect Science Lessons* (2007)

Figure 1

Raised-bed garden

The project was based around specific science standards. In addition students were asked to research using nonfiction materials; create detailed summaries of all garden elements; work collaboratively with fellow students to design, plant, and present the garden; communicate with adults in the community; design and construct the garden using concepts of measurement and geometry; and create interpretative drawings of garden elements using a variety of artistic media.

By the completion of the project, students were able to clearly articulate what plants need to survive and how plants have been used in our area throughout history.

DESIGNING THE GARDEN

Our first planning hurdle was how we could design the project to give us the biggest educational bang for our buck, in this case $25. Because of other learning activities already planned with both of our budgets, my fellow teacher and I could only allot this amount (total) for the garden project. We decided to make this budgetary limitation part of the planning process for students, giving them an added lesson in economics!

We started by dividing the 40 ft. × 15 ft. area between our portables into three smaller sections. I then assigned each student one section for which they were to design a raised-bed layout on graph paper using their understanding of area and perimeter. The beds were to be framed with lumber pieces to make garden boxes. Students' designs had to follow the provided scale, be measured accurately, and include maneuverable pathways and safe distances from surrounding structures—the back of an adjacent portable building and the porch area of our portable classrooms. (School district maintenance workers needed to have access to all parts of the portables in case of repair or improvement, and we did not want our garden damaged because we did not follow the necessary measuring requirements.)

Because our garden area was between two buildings, it was partially shaded and relatively small in comparison to our other school gardens. As a result, students had to choose native plants that could grow in a small, partially shaded environment, which for this garden meant plants such as ground covers, shrubs, and delicate tree varieties. (Larger trees would not have enough space or sunlight to maintain a healthy life and a strong root system could cause damage to the surrounding portable buildings.)

Completed designs were voted on by the class; three plans were chosen—one for each section of the garden space:

- Two long, skinny beds capable of holding several shrubs and ground covers;
- A series of six smaller-size beds that could hold one to four plants each, depending on the amount of growth space needed by each plant; or
- Two large, square-shape beds that allowed for ultimate planting flexibility and the ability to accommodate a wide range of native plants.

Figure 2

Native plant list with ethnobotanical uses

Salal: food source, hunger suppressant
Red Osier Dogwood: dyes, food source, fish traps
Oceanspray: medicinal (eye wash, diarrhea, small pox)
False Lily of the Valley: medicinal (heart problems)
Oregon Grape: food source, dyes
Sword Fern: flooring, bedding, food preparation
Bracken Fern: food source
Lady Fern: food source, medicinal tea
Goat's Beard: medicinal (liver and gallbladder), skin treatment
Salmonberry: food source
Bleeding Heart: medicinal tonic
Red Flowering Currant: food source
Serviceberry: food source
Bluebell: glue, laundering, ceremonial
Red Huckleberry: food source, medicinal (sore throat), fish bait
Vine Maple: fishing nets, snow shoes, baskets

In all three cases, the space left for the pathway was adequate and worked well with the other chosen designs. To make the project more realistic for our budget, we asked students to choose the *one* garden section they thought would make the most impact to build this year, and the other two sections would be developed in later years or as funding or materials became available. They decided they wanted to build the garden bed with six 4 ft. × 3 ft. boxes containing 17 different varieties of native plants.

BUILDING THE GARDEN

Luckily for us, some spare lumber from a previous project on school grounds was available to frame the bed boxes, so with the boxes soon ready to go, we needed only soil, materials for a pathway, and plants. The students came up with the idea of asking businesses to donate the remaining materials needed. To save time, we asked parent volunteers from our PTSA to help us solicit donations and look for additional resources. Word of mouth went a long way in helping us accomplish this project; local businesses and community members seemed more than happy to help when they heard what we were trying to do. For example, Lowe's donated several bags of soil, and a local nursery and community field biologist donated some native plants. (We also purchased plants ourselves, spending our $25 at a local plant sale to purchase additional native varieties.)

Using tools brought from home, students spent every afternoon science period and recess time for the next five weeks putting the garden together. We provided guiding questions about various garden elements for the students to research in small groups and determine the best way to incorporate them into our garden. Tasks included everything from leveling the ground and problem solving erosion with a drainage ditch to framing the garden boxes, filling them with soil, and adding the chosen plants. Know the source of the soils you work with. Make sure students wash hands thoroughly after handling soil and plants. When using gardening tools, make sure students handle tools properly and wear heavy duty gardening gloves to avoid splinters.

Students also decided the garden would look more natural if they included items from the surrounding area on the pathways, such as sticks, pinecones, pine needles, and stones of all sizes.

After the garden was built, students typed and laminated plant identification tags and signs describing the various processes undertaken in creating the garden. Students also created signs describing the box placement, pathway construction, the purpose of the drainage ditch and rain barrel, and the different types of mulch used in each plant box.

NATIVE PLANT LEARNING

While students were building the garden outside, they were learning about the purpose and importance of native plants inside. As the garden project got under way, we had lots of discussion about native plants and provided students with an extensive list of native plants from our region. Students then researched the plants on the internet and in native plant guides to choose the ones they were most interested in planting in our garden. Because we wanted this part of the project to be as student-led as possible, we left it up to the students to find plants that fit our garden's criteria—partial shade, small growing space, temperate climate. An additional requirement—per our request—was that each plant have an *enthnobotanical* use (meaning a human use, be it medicinal, culinary, or otherwise). Students researched these elements using online state-specific native plant websites and local field guides.

This helped us tie the experience into our social studies curriculum. Prior to this project, students had learned about the Lewis and Clark expedition and the Oregon Trail. During these studies, we introduced the role that native plants had played in the success of these trips. We also discussed how the Native Americans from

our area used these plants to ensure their own survival. When the gardening project was under way, student research uncovered several different uses for various plants, including food, healing, hunger suppression, and dyes. Discovering these alternative uses helped students better understand the importance of growing and using native plants, even today.

ASSESSMENTS

Throughout the project, we assessed student progress in several ways. For example, each student was responsible for researching one garden-related topic (assigned by the teachers) and summarizing it. In grading the writing, we looked for accuracy of the content and ability to explain ideas clearly. Garden layouts were also assessed for mathematical understanding, and each student received a participation grade based on the amount of effort they put toward helping complete the garden. At the culmination of the project, students also completed self-reflections that allowed them to share their thoughts about the garden project, including likes, dislikes, and overall learning.

To celebrate our achievements with the community at large, students planned and hosted a ribbon-cutting ceremony for parents and other community members. Students auditioned for the speaking roles and were chosen by a voting process. Those not chosen to speak became ribbon cutters, food servers, and greeters. All of the students worked together to prepare food items and decorate the classroom. Students also wrote letters to the local television stations and newspapers inviting them to our event. On ceremony day, every student beamed with pride as they looked over the garden and realized they had made something amazing!

As a culminating effort, we compiled their summaries into a book, adding pictures of the process, feedback from garden visitors, and the newspaper article that was written about the garden. This book is evidence of student learning. It demonstrates student understanding of each section of the garden, as well as writing ability.

ANYTHING IS POSSIBLE

In the end, the lessons learned for all of us—both students and teachers—were innumerable. Students gained valuable academic knowledge and practical gardening skills, and they also developed abilities to solve problems and work together as a team. We learned that money does not have to be a barrier to making your ideas a reality. With creative thinking and scavenging, anything is possible. PTSA, community members, and local businesses value education and want to help enhance student learning in whatever way they can. It never hurts to ask for their help, especially because the answer will often be "yes."

Acknowledgment
Thank you to fellow fourth-grade teacher Sue Macnamara for helping me make our native plant learning garden a reality.

Reference
National Research Council (NRC). 1996. *National science education standards.* Washington, DC: National Academies Press.

Internet Resources
Create a Certified Wildlife Habitat
www.nwf.org/backyard
Discovery Elementary Native Plant Learning Garden
www.discovery.issaquah.wednet.edu/Outdoor%20 Classrooms/Native%20Plant.htm
eNature.com
www.enature.com/home/indexNew.asp
Kids Gardening, National Gardening Association
www.kidsgardening.org
National Gardening Association
www.garden.org

This article first appeared in the Summer 2008 issue of Science and Children.

Chapter 29

Sand

Up Close and Amazing

by Thomas E. McDuffie

We walk and ride on it often, slide on it playing softball, or collide with it during football season. Some like to sleep on it. Forensic scientists examine it for crime clues. Engineers test its strength when cemented together; wood workers use it as an abrasive; and Mother Nature mixes it with organic matter to grow our food. What is it? Sand, of course! We take these tiny bits of rock for granted, and sometimes even consider them a nuisance.

My interest in sand began out of boredom and involved little more than sifting it through my fingers. I first noticed the sparkle; then the variety of colors, sizes, and shapes caught my attention. Over time, speculation about the probable origin of a wide range of sands simply became fun. From this interest in sand I developed and field-tested a series of activities with middle level teachers and students (though the activities are adaptable for use with younger or older students as well). When complemented with the extension ideas, these hands-on activities are interdisciplinary, capture the essence of the Standards (properties of Earth materials and changes in properties as well as interactions within systems), address varied instructional approaches, and incorporate technology to explore sands from around the world.

NATURE OF SAND

What is sand? How is it formed? Do sands differ from location to location? What causes the varied colors, textures, shapes, and sizes of sand particles? How is sand different from soil? Sand is defined as any natural mineral material with grain sizes between 0.02 and 2.0 mm in diameter. Gravel has larger grains, while those found in silts and clays are smaller.

Sand (like gravel, silt, and clay) is sediment created by the mechanical and chemical breakdown of parent, or source, rocks. Outcroppings, river valleys, and road cuts provide good sites to observe the results of the breakdown process. Particles freshly broken from the parent rock resemble the source in terms of color and mineral content. During the usually lengthy, gravity-driven journey from source to final resting place, the particles of rock are altered mechanically when

Further Reading

- "School Yard Geology," from the December 2008 issue of *Science Scope*
- "My Favorite Rock," from *A Head Start on Science* (2007)

rock fragments bounce or grind against one another causing breaking, chipping, and flaking. Chemical breakdown occurs because rain and groundwater are naturally weak acids (which can become stronger with pollution) that dissolve components of rock such as calcium carbonate and mica.

Throughout the eons-long process of weathering and erosion by air, water, and ice, sand particles are precipitated by moving water in or near rivers, lakes, or the ocean to form deposits as natural levees, sand bars, beaches, and dunes. If they are carried by winds, grains bounce along and occasionally crash into one another to make finer, smoother particles. Ultimately this can form the huge, shifting dunes found in deserts or along current and former ocean margins (see Internet Resources for information about the special case of White Sands, New Mexico). The finest of particles from the Sahara desert—actually the size of silt or clay—are swept high into the atmosphere and can travel thousands of miles before precipitating (the rich soil in Bermuda can be traced to this phenomenon).

The transportation process is largely responsible for the size, shape, sorting, and texture of sediments. Given similar source rocks, small, round particles travel farther than larger, angular ones because of the extended erosion period. Similarly, the composition of sand and its color remain linked to the source rock. The black, angular, shiny sand found on some Hawaiian beaches is composed of volcanic obsidian fragments. The light pink to white sands in the Caribbean are predominantly ground shells and coral. By contrast, the sediment forming Long Island beaches is largely the result of the breakdown of New England granite relocated by glaciers—quartz, feldspar, and some mica mixed with shell fragments.

The origins of most sands are not as easily determined as these examples. Throughout the erosion process, agents such as acidic rainfall change sand particles chemically. Nonetheless, students can learn a great deal about sand by examining the composition, size, rounding, and sorting of grains.

DEVELOPING A SAND COLLECTION

To make the study of sands come alive, it helps to develop a collection from various environments—river, lake, and ocean beaches, as well as deserts, mountains, and fields. This allows students to make comparisons among a variety of sources. The first time I tried this activity, I had only a few samples collected from my backyard, a local stream, a building site, etc., but now my sand collection includes specimens from around the country and world. When I travel, I carry film

canisters or closeable plastic sandwich bags to collect sand because they are both easily labeled and stored. Yet, students and friends who travel on business or vacations have contributed the most samples. A pen-pal option can also be used to contact sand collectors and swap samples (see Internet Resources).

Preparing samples for class use requires minimal effort. Washing samples several times with tap water to separate organic matter and fine particles—silts and clay—is the main step. Pouring the organic material and dirty water into a plastic bucket rather than the sink prevents clogs, and the buckets can be emptied outside. Clean, dried specimens are then stored in jars and labeled by location for easy observation. If you do not have a substantial sand collection immediately available for this activity, use sand resources provided as a source of virtual information that illustrates the same content. At the very least, a virtual tour can demonstrate the similarities and differences of sands from around the world. Moreover, these locations can be plotted on a map of the country or world to emphasize geographic skills.

STUDENT INVESTIGATIONS

The activities included are structured around the collecting, cleaning, and comparing of sands from a range of locations and are enhanced by various extension ideas. I have designed the activities to include some at-home phases, however, the studies can just as easily be done entirely in the classroom and around the school.

In Activity 1, the study of sand begins with each student collecting a sample from home. To make the topic more meaningful, and to reinforce the idea of place, students should collect samples locally, even if it means a trip to the nearest park. A surprisingly small sample, one-quarter of a baby food jar or a corner of a sandwich bag, is enough to begin a permanent collection. Students should record the location where the sample was collected.

Unless your school is near a coast or dune area, most students are likely to collect soil. Soil is formed from rock particles and decaying plants and animals, and sand is only part of it. Different-size mineral particles give soil its texture. Organic matter helps provide color and nutrients. Although our principal focus is sand, soil is too important to disregard. To extend the study of soil, students can read about it in an encyclopedia or in *Handful of Dirt* (see Resources) or explore its types and variety online (see Internet Resources).

Back in class with their samples, students are ready to begin Activity 2. To observe grains clearly,

Activity Sheet 1

Collecting samples at home

Materials
- Baby food jars or resealable sandwich bags
- Magnifying glasses
- Piece of white paper

Procedure
1. Before you do any gathering
   - Predict what the samples will look like.
   - Think about what the sample might contain.
2. Fill one-quarter of a baby food jar or a corner of a sandwich bag with sand.
3. Record the location where the sample was collected.
4. Examine a very small amount of the material—tip of a teaspoon—using a magnifying glass.
5. Draw and describe what you see in the sample.

Questions
1. What does the sample contain? Why can it be considered a mixture?
2. Do you see any sand grains? Draw them; then use adjectives to describe them.
3. Do you find any evidence of plant life? Draw it; then compare its appearance with sand.
4. Do you find any man-made material? Why do you think it was made?

the samples must be washed several times and then dried thoroughly. Because fine-grained samples can take days to dry, washing is best done before a weekend. Working in cooperative groups of two or three, students are given an overview of the lab activities; directed to cover the work areas with newspaper; and cautioned to use a minimal amount of water for cleaning, to shake the mixture gently, and to repeat the rinsing process several times. After samples are labeled, organized by groups, and stored either on a table or window ledge to dry, we begin to examine the leftover organic matter.

Cleaning the samples should allow plenty of time to examine the organic components that float to the surface. Students use several senses in examining the organic component (sight, smell, and touch) and answer the review questions as a class. Why did the soil separate? What did you see in the part that floats? What is this part made up of? What causes parts to sink or float? A discussion of the soil's composition, the effects of water, fertile or nonfertile soils, and making soil can occur if time permits. The idea that soil

can be separated through flotation and the composition of the organic component surprises many students. Of course "middle schoolers" will let you know

Activity Sheet 2

Cleaning samples

Materials
- Baby food jars
- Trays or paper plates
- Resealable sandwich bags
- Paper towels
- Water
- Magnifying glasses
- Plastic spoons
- Pencils
- Waste container

Procedure
1. Wash the samples several times, then dry thoroughly.
2. Place two or three spoonfuls of the sample in a baby food jar or sandwich bag.
3. Cover with water and shake gently.
4. With the plastic spoon, scoop out some of the debris that floated on the water and place it on a piece of paper towel. This material will be examined in a few minutes.
5. Pour the excess water and debris into the waste container to avoid clogging the sink.
6. Repeat the addition of water, shaking, and pouring until excess liquid is clear.
7. Put a piece of paper towel on a paper plate or tray.
8. Pour all excess water out, then dump sand onto the paper towel and pat gently.
9. Place each group member's sample on a tray or paper plate. Label it and put it in the designated location for drying.
10. Get the material set aside in step four. Spread it out with the tip of a pencil.
11. Examine the contents, then answer the follow-up questions.

Questions
1. Why do the samples need to be washed?
2. What happened to the sample during the first cleaning? Why did some of it float?
3. After observing the substance with a magnifying glass, describe the material you scooped out.
4. How are the materials that floated and sank similar and different?
5. How can soil be made?

that it's a good idea to clean hands after touching the decomposing litter.

Since the actual examination of sands is fundamentally an observational activity it is, in theory, easily performed. In practice it is beneficial for students to observe two of the teacher's sand samples with different characteristics both to practice procedures and to establish working definitions of color, size, shape, and texture; the chart can be completed as a group for these samples. Next, students collect and record data for their groups' samples. Beyond the data categories, teachers can easily promote observation skills with guiding questions: How are the samples the same? How are they different? How far apart are the sites where the samples were taken? What could cause the differences? Time should permit sharing sample "slides" with one or two other groups so students can observe about a dozen different samples. Because samples from the same locale may exhibit a great deal of similarity in basic characteristics, prod students to make precise observations: Is there a difference between samples from different parts of the community?

Your sand collection and access to computers dictates the approach you can take on identifying sands and the broader issues, characteristics, and processes of sand. With my teaching collection, I give students a crime scene investigation scenario. Naturally the "unknown" sample each group must identify comes from the shoes of a suspect for some heinous crime! Their job is to find the sand's original location based on its properties. With 8–10 student groups, I use no more than 15 samples chosen for their distinct characteristics. To facilitate the identification process, at least three sets of samples and sample keys must be available for student comparisons. The samples should exhibit the characteristics of different geologic features—beaches, glacial deposits, wind-blown dunes, lake and river bottoms—or contain a range of minerals or color variations.

Concepts such as weathering and erosion, transport, sorting and deposition, and sand's effect on rock particles can be introduced or reinforced through these activities. Moreover, the texture of sand particles can be related to land features such as moraines, dunes, or beaches, while the grain size can be related to transportation distance. Human use of sand, from making glass to making cement, can also be introduced. Finally, I like to share unusual, motivating samples such as "magnetic" sand that contains magnetite, samples with clear crystalline structures, beach sand made of black glass, beaches with no sand, and clear glass sand.

In a real sense, the sand activities provide a vehicle that introduces students to the changes rock undergoes as it weathers, erodes, and is deposited. Ultimately an outcropping is broken down and moved toward the sea. Along the way it is changed physically and chemically, sorted by size, and becomes part of soil or is used for a variety of products.

EXTENSION ACTIVITIES

Given the many opportunities for extension, time is the limiting factor. The reading-writing connection is a natural. Students can not only read stories (see Resources) but also can create their own fiction about the making or movement of sand. Additional writing possibilities include research projects into sand-related products (e.g., glass, bricks, or petroleum sand), or sand castle art projects. Sand paintings can be a portal into Navajo spirituality. One of my favorite ways of expanding the study of sand and soil involves the making of adobe bricks.

CONCLUSION

The integration of science with social science and literature captures the spirit of inquiry and pedagogy embodied in the National Science Education Standards. Multiple instructional approaches—group and individual work; small and large group work; at-home activities; hands-on and virtual instruction; extensions into literature, writing, and geography—meet the needs of students with different learning styles. The sand activities potentially link hands-on instruction with distance learning through the websites referenced. Lastly, these inexpensive and easily performed activities provide a conceptual introduction to weathering, erosion, and deposition as part of the rock cycle. They also provide an inviting look at the often-unnoticed world beneath our feet.

Resources

Bial, R. 2000. *A handful of dirt*. New York: Walker.
The award-winning, beautifully illustrated book invites exploration of soil and the life it supports by youngsters (grades 4–8).
FOSS: Full Option Science System. 1995. *Pebbles, sand and silt*. Nashua, NH: Delta Education.
This unit is part of the major FOSS curriculum project. It provides a range of hands-on activities that can be readily adapted for middle-grade youngsters.
Gallant, R. A. 1997. *Sand on the move*. New York: Franklin Watts.

NATIONAL SCIENCE TEACHERS ASSOCIATION

Activity Sheet 3

Comparing and identifying samples

Materials
- The display collection representing a broad range of sand types and locations
- 8–10 different samples from varied locals
- Student samples
- Internet access
- Transparent tape
- White piece of paper
- Magnifying glasses or handheld microscopes

Part 1: Comparing Student Samples
Procedure
1. Wrap a piece of transparent tape 6–8 cm long over the eraser end of a pencil and up the sides with the adhesive side out.
2. Dip the transparent tape into the sand sample using the adhesive side to pick up the grains.
3. Place the tape's adhesive side onto a piece of paper to make a tape slide.
4. Examine the sand using a magnifying glass or handheld microscope.
5. Compare your sample with samples collected by other students, and fill in the data chart below.

Questions
1. Are grains from most of the samples the same color? How are the grains similar/different? Why?
2. What do you think influenced the size and shape of the grains?
3. How does the source influence the characteristics of sands?

Sample number/name	Color(s)	Shape of particles	Sorting/size of particles	Clarity (translucent, transparent, or opaque)	Luster (shiny or dull)	Rough or smooth

Part 2: Identifying Unknown Sand Sample
Procedure
1. Using samples of six different types of sand, create tape slides as you did in Part 1 of the activity.
2. Examine the sand using a magnifying glass or handheld microscope.
3. With the data chart based on your observations as your guide, use the internet to compare your sample with those collected at three places in the world.
4. Find the sources of the virtual samples on a map and record their latitude and longitude.

Questions
1. How are local sands similar and different?
2. What might cause their color? Hint: Sand contains minerals that include various chemicals.
3. How do local sands compare with those in the locations you selected?

The author of many science books presents the formation, shapes, and movement of sand dunes and the biotic communities they contain (grades 4–7).

Prager, E. J. 2000. *Sand.* Washington, DC: National Geographic Society.

This easily read (grades 3–5), highly illustrated children's book realistically and accurately describes the formation, movement, and color of sands.

Internet Resources

White Sands National Monument
www.nps.gov/whsa/dunes.htm

Soil Science Testing
www.pedosphere.com/interactive.cfm?rgnum=29140

Oil Sands Discovery Center
www.oilsandsdiscovery.com

Sand Art Activity
http://crafts.kaboose.com/sand-art.html

The Virtual Sandbox Museum of World Sands
www.jaster.20m.com/index.html

This article first appeared in the September 2003 issue of Science Scope.

PART

5

Funds and Materials

Chapter 30

You *Can* Get What You Want

Tried-and-True Tips for Securing Funds and Resources From the Community

by Yvonne Delgado

Many teachers regularly use their own money during the year to pay for miscellaneous teaching materials or to help individual students in need. What if I told you that you could secure all the funds and resources for anything that you want to do educationally for your classroom? It's no fantasy; it's simply a matter of thinking ahead and expending some up-front energy for some great returns on the time investment.

COMMON MATERIALS

Let's first talk about securing common materials or supplies. Let's say you need clear plastic pop bottles to fill with water to make Cartesian divers for science. Possibly you're even getting sick of drinking all that pop, rinsing bottles, and carrying them to school. What if you could get all the clear plastic water bottles you needed—clean bottles, bottles with lids, and bottles already full of water. This would be an easy task. Look in the phone book. Find some water companies or grocery stores in the local vicinity. If you need just a few bottles, you can ask any local company. The more you need, the more you may want to approach a larger organization, and the more closely you would want to match your request with their product line. Call first; let them know you're looking for water

bottle donations. Find out the name of the person who handles such requests. Know exactly what you want (specific size, quantity, shape, color) before calling, and then ask for it. I shoot for the ideal, and most of the time I get what I ask for.

Be prepared with the following information:
- Your name, title, and connection to the project
- Name of school, location, contact information
- Number of classrooms and students being served
- In some instances, demographics are also helpful (e.g., percentage of students not graduating, number of minority students, number of students who receive free and reduced lunch, number of students by gender)
- Subject area
- General age range of students
- Exact item specifications and quantity needed
- Date items are needed (give yourself some leeway; don't cut it too close)
- What is in it for the company or organization (e.g., good public relations)

Steps in securing donations:
- List needs and wants prior to the start of the school year or the project start date.
- Gather all information.

Examples of Donations

Here is a sampling of the donations of resources and time I have secured over the years:

- College professors providing professional development for our teaching staff
- Lunches from local restaurants for a summer educational program
- Donations for a basket to be auctioned off at a charitable event
- Dry ice for science experiments
- Labor for installing new carpeting and painting classrooms
- Computers for teaching staff
- Laptop and projector for teaching
- Three years of maintenance of our microscopes
- Paper, markers, paints, cardboard, cups, etc., for various science events
- Baking soda, vinegar, candies, sugar, salt, etc., for various science experiments
- Local scientists, artists, actors, and authors to provide authenticity to a subject (either as a guest speaker, to assist with a project, or to demonstrate their craft)
- Time to judge a science event
- Gift certificates as special prizes for students
- Books on a specific subject for students

- Be specific on exactly what you want (e.g., five 3' × 5' pieces of sod).
- Try to match requests with company product lines.
- Be prompt picking up items.
- Send a timely thank you (whenever appropriate, I promote their donation in flyers, news releases, or publications).

Some examples of commonly requested items are batteries, spools of wire, fabric, special types of paper, pens or markers, paper cups—the list is endless. As for quantity, ask for what you need. If you need only a couple items because you are demonstrating something, then ask only for that. If you need enough for a group of students, ask for enough for each student plus a couple extra. The companies will do their best to meet your request or assist as much as they can. Once I needed candy for a mathematics sorting project. I wrote a letter explaining the information as outlined on this page and requested five bags of a specific type of candy. I stopped in, talked to the manager, explained my request, and handed him my letter. The next day he called and donated a $25 gift card from the store, which more than satisfied my candy purchase.

PROFESSIONAL VOLUNTEERS

Another type of need is to find people to construct larger or more complex project components. Again, use your local resources and collect community information. As you drive around your community, collect the names of the local companies you pass on the way. When you talk to parents, make note of what organizations they work for. Read the local papers; find out who's who in the business community. These are the types of long-term relationships that you want to cultivate. Keep the company names, individual contacts, and phone numbers in your computer, PDA, or phone.

When I embark on an environmental project, for example, I call in volunteers from the local parks, the county soil and water department, and environmental agencies. I also contact specialists from local businesses and other subject matter experts. If I'm working on an autism project, I contact school special education staff, local autism nonprofits, parents with autistic children, and so on. It has been my experience that most everyone is willing and able to assist if given enough notice. I would recommend at least one month of lead time to secure a volunteer.

Working with a team allows the responsibilities to be split into reasonable workloads. Recently I worked on a project to update and enhance our existing erosion lesson. I called and secured a volunteer from each of the following organizations—the local park system, the county soil and water department, and one other local science education specialist. I then set a series of team meetings—one hour every two weeks for a period of three months. We selected the meeting time to best meet the needs of the team players. Two teachers were the team leaders and met with the team every two weeks. At the first meeting, we provided the team with a general overview and project expectations—complete with a timeline and any budgetary information. These local subject matter experts took our teachers out to sites to see real examples of erosion. In addition, they were able to share valuable information about the various types of erosion, what additional problems arise with erosion, and contemporary erosion control practices. The volunteers assisted with building some of the erosion demonstration models, donated literature and other materials for the classroom, gave examples of how they teach the topic to students, and overall provided our teachers with a great educational experience that could be shared with the students.

Another option to consider is contacting local colleges, vocational schools, or trade schools. Often teachers in these institutions are looking for opportunities to engage their students in real projects. For example, after receiving funds from a grant to purchase wiring, I asked electrical students from a vocational school to come in and rewire our program's computers. It was a win-win situation for all involved.

Points to consider when asking for volunteers:

- Gather your information.
- Be specific about the project needs and parameters.
- Be sure to let them know what their commitment involves: number of meetings, approximate amount of time at each meeting, deadline for the project.
- Let them know where the funds are coming for the materials and supplies.
- Be sensitive to their time needs—keep meetings on schedule and limit meetings to only what is needed.
- Always remember to send a thank-you note upon project completion or once you have received your items. I sometimes include a photograph or have students sign the thank-you note. Often, these professionals are flattered that you recognized them and their ability to contribute to an important educational project.

MORE COMPLEX PROJECTS

Donations have enhanced our science program in many ways, but two of my favorite examples demonstrate how such contributions can make a huge difference in your science curriculum.

When our science department decided to build erosion tables to use for hands-on Earth science investigations of erosion, we needed sod and sand. So I identified companies who specialized in each of these areas—a nursery and a sand company. Our tables have three columns each—one holds sod and two hold sand. As we gathered information for the nursery, we had to measure each column to determine the exact dimensions. With the dimensions in hand, we approached the nursery and were provided with a bundle of sod that we cut to fit the columns. The cutting was more difficult than we anticipated, so the following year we asked nursery staff to assist us with cutting the sod into the proper dimensions. For the sand donation, we needed to specify the grade and quantity. We first had to experiment to get the proper grade that would best demonstrate erosion, which we did. The volunteers so enjoyed working on creating these indoor

Connecting to the Standards

This article relates to the following *National Science Education Standards* (NRC 1996).

Teaching Standards

Standard F:
Teachers of science actively participate in the ongoing planning and development of the school science program.

Program Standards

Standard D:
The K–12 science program must give students access to appropriate and sufficient resources, including quality teachers, time, materials and equipment, adequate and safe space, and the community.

environmental learning activities, we are now working together to design an outdoor component.

Another special feature of our science classroom made possible through donations are two 180 gal. aquariums provided to us through a grant from the Hershey Foundation. One aquarium houses native Ohio pond fish and animals. The other is home to native Ohio stream fish and animals. The stream tank also has a chiller and flow system to best simulate the natural environment of a stream. We use these tanks to teach classification, habitat, and the water cycle. To help offset the costs of maintaining the tanks, local bait shops were approached for donations of feed—which they gladly gave. We continue to value each and every contribution to our many projects.

Although it can seem impossible to accomplish some of the projects we dream of doing with our students, with a little help, you can do a lot. So the next time a blockbuster idea for a new learning project strikes, instead of feeling limited by a lack of dollars or supplies, grab a list of Who's Who from the local Chamber of Commerce. Remember, you'll never know unless you ask!

Reference

National Research Council (NRC). 1996. *National science education standards.* Washington, DC: National Academies Press.

This article first appeared in the Summer 2008 issue of Science and Children.

Chapter 31

Need Money? Get a Grant!

Tips on Writing Grants for Classroom Materials and Larger Items

By Linda Bryson

They say necessity is the mother of invention. I guess that's how I became my school's resident grant writer. It's not that I had particular gifts as a writer or an uncanny sense of persuasion or even a special gift of gab. What I had was a need, and the need was for money! The first grant I received was a Lysol/NSTA award in 2001. That grant gave me $1,000 to spend on registration, travel, and housing so I could attend NSTA's National Conference in San Diego, California, as well as $500 to purchase classroom supplies. It wasn't a lot of money, but it did inspire me to try my hand at applying for additional grants. Once I got started, I couldn't stop. After I obtained countless grants and brought more than $20,000 in money and materials to my school as a result of these efforts, people started asking *me* for help getting grants. So I've pulled together a few of my favorite tips to help you get started finding that most elusive of resource—money—for your classroom. If I could only figure out how to get some more time. . . .

WORK THE INTERNET

The internet is one of the easiest ways to locate potential grants. Typing in a general search for "teacher grant money" gets you started. Different search engines give different results. For example, Google results differ

from Yahoo! results, which differ from those found on Mamma.com (I usually have good luck with Mamma). This may seem strange, but I sometimes get different results for the same search at different times during the day, so be sure to try different times as well.

As soon as I get the search results, I e-mail them to myself. Believe me, it's easier to delete an e-mail link that you are no longer interested in than to feel like kicking yourself for not being able to find the link again.

I start by eliminating the grants I don't qualify for by looking at grade level and state requirements under "eligibility requirements." A grant may sound great, but it may not be appropriate for you, so don't waste any of your valuable time.

If you do qualify by grade level and state, then give every grant a chance. The title of a grant could make you think that it isn't for you, but if you look closely, there may be something there. For example, I learned of the Lysol/NSTA grant from an e-mail from NSTA that contained information about the elementary grant. The title Lysol/NSTA didn't initially spark any connections to my curriculum. However, I read the grant application carefully and found that it was something that interested me. While the majority of the winners had projects about sneezing, blowing their noses, and germs, my winning project was on monarch butterflies. I had read the grant application's fine print, which said the proposal could be on any topic. And my project was something that I had completed previously with my students, which made it easy for me to write it up at a busy time of the year—another useful tip to file away when considering applying for grants.

GATHER SCHOOL DATA

Once you decide to apply for a grant for which you are qualified, carefully read the fine print to make sure you meet all of the requirements. For this part, you will probably also need to know some basic information about your school, such as the population breakdown by ethnicity or maybe the percentage of students who receive free lunches. This information is available through your state's department of education. Unfortunately, this information may be more than a year old. Your current school's population breakdown is available through your district's central office, and you may have to ask around to find someone who knows the current data. Fortunately, computers have made this information available at a person's fingertips.

STORE YOUR STUFF

Once you have some basic information, you may need it for another grant application, so store it in a special folder or notebook. I use a pocket folder with brad fasteners. Any grant-related information that I need to keep for future reference gets holes punched into it and fastened into the middle section of the folder. Any completed grant applications also get added to the middle section. I use the folder's pockets for other grant-related items—one pocket for grants that I might be considering and the other for incomplete applications.

If you print out a grant application with several pages, make sure to staple it together before you insert it into a pocket. It is easier to have the pages all together when you are looking for it again at a later time. As time goes by, periodically clean out your grants folder to keep it up-to-date.

GIVE THE DATE ITS DUE

A proposal's due date is probably the most important piece of information in the application process. It may be a busy time of the year, so you have to realistically decide if it is something that you are going to be able to complete before the due date. I submitted one grant application in a 48-hour period of time, and I could do it because I had all of my information already organized. Still, I was shocked when I found out that I

had won the grant, because I had completed the actual write-up in less than two hours.

COMPLETE THE APPLICATION

Once you begin filling out your grant application, make sure to complete all sections. Read and then reread the questions to make sure you understand what information is required. Some grants have links to frequently asked questions where the organization has already taken the time to give a more thorough explanation to some basic questions.

If you can, work with a peer or friendly critic, then have him or her read over the application to make sure that your answers are clear and appropriate.

If there is a section marked "optional," complete that, too. Even if it says optional, I don't take any chances and supply the information.

For some grants, you must send a letter of intent before receiving the actual grant application. If this is a requirement, write up your letter, show it to your administrator, and then ask if it can be copied onto your school's letterhead. The school letterhead gives more credence to your request.

If the application is sent via the internet, sometimes it is a good idea to print out the application first to view the overall grant requirements in one place. This will also allow you a chance to gather any necessary information before you actually sit down and begin filling out the application. Use your folder or notebook and have all of the needed information at your fingertips, so you don't have to start and stop or get up because you need to find another piece of information. I also find that using a highlighter to mark important information is helpful. Printing out the confirmation sheet that you receive after you send the application provides a reminder for later.

LOOK IN MANY PLACES

There is no one single place for grants—grants are everywhere, even places you might not think. Grants for specialized areas of interest are also often sent through e-mails. Join an organization, and get on their e-mail lists. As a member of NSTA and NSTA's Building a Presence, I receive e-mails about numerous opportunities. Even if it is not appropriate for me or my school, it only takes a minute to forward an e-mail to someone who may find the opportunity of interest. That road goes two ways—if you pass along opportunities that may interest others and let the word out that you are interested in applying for grant money, they in turn may be likely to pass along opportunities they come across that are appropriate for you.

Many national grants may come from unexpected places. For example, national retailers, such as Best Buy, Verizon, and Target may have educational grants on their websites. Check them out. (See Figure 1 for more grant ideas.)

SPREAD THE WEALTH

I have written many grants in the same school year for similar materials. This past year, I wrote several grants for computer projection equipment. I started off by having just my name on the first grant application. As the school year went by, I asked other staff members if I could include their names on the grant. That way, if I receive a grant, other teachers in my building will benefit if I get another one. I give duplicate materials to someone else in our building. So far, all of my colleagues have been happy to be added to the grant applications.

Working with other supportive technology staff members, if you need some guidance, is also helpful. On one grant for classroom computer equipment, I included the librarian's name in the grant, but by working with our district's technology staff, we ended up equipping my classroom *and* the library. That made a lot of us happy.

I also tear out clippings about grants from professional newspapers and magazines. They go into the folder's pocket until I have the time to read them over.

Connecting to the Standards

This article relates to the following *National Science Education Standards* (NRC 1996):

Teaching Standards
Standard D:
Teachers of science design and manage learning environments that provide students with the time, space, and resources needed for learning science.

BRACE FOR SUCCESS!

Once people in your area see that you are successful at grant writing, they will want to talk to you. I

Figure 1

A selection of helpful grant resources

Grant Resources From NSTA
Grants, Grants, Grants
http://sciguides.nsta.org/internet/grants.aspx
NSTA's Webwatchers program has compiled a list of online resources to help you find grant money and to help you get it.

NSTA Teacher Awards and Competitions
www.nsta.org/awardscomp
A listing of other awards and opportunities available through NSTA

Other Grant Opportunities
Best Buy Teach Awards
http://communications.bestbuy.com/ communityrelations/teach.asp

Target Field Trip Grants
http://sites.target.com/site/en/ corporate/page.jsp?contentId =PRD03-002537

Verizon Foundation
http://foundation.verizon.com
Click on "apply for grants" to learn more.

have worked with middle and high school teachers in our area on grant writing. I give a short introduction to get them started and then turn them loose in a computer lab. They work for hours and say very little because everyone is interested in finding money to spend in their classrooms! With a little perseverance, they'll find it!

Resource

National Research Council (NRC). 1996. *National science education standards.* Washington, DC: National Academies Press.

This article first appeared in the Summer 2007 issue of Science and Children.

Chapter 32

Science on a Shoestring

Stock Your Shelves With Free and Inexpensive Science Materials

by Sandy Watson

Most of us have experienced the frustration of limited school science budgets, and many of us have had to resort to repeatedly dipping into our personal funds to finance the material needs of our classrooms. Certainly there are some items, such as chemicals and safety equipment, that must be ordered from educational science supply companies, but there are many other items that can be acquired at little or no cost to the teacher or school system. With a little time and effort, you can start the school year with adequately stocked science supply shelves.

Before beginning your search for materials, check to see if your school has a policy governing the use of items that have been donated or obtained from nontraditional sources (such as local businesses). Also, double-check your budget for the year and review your school's reimbursement policy before spending any of your own money. Also, take a look through the laboratory activities you have done in the past and want to repeat, and create a list of materials for each activity. If you are looking for low-cost labs to substitute for more expensive activities, Figure 1, page 154, lists websites where labs that use inexpensive materials can be found. Peruse those sites and see if anything appeals to you. Save the labs that you feel you could duplicate and add the necessary materials to your supply list. Figure 2 on page 154 is a sample list of inexpensive

and free science supplies that you can send home to parents with a request for donations. This list can also be posted in the teachers' lounge.

Great places to begin your search for classroom supplies are yard sales, garage sales, and flea markets. Be sure to take your list with you and be prepared to dig through boxes. Some people will place a box of toys out and it will be up to you to search through it for balls, marbles, small cars, and other items on your list. Once they learn you are a teacher and are looking for items to use with your students, many people will discount the price of some items or even donate them. I once obtained a large box of more than 150 brand-new amber dropper bottles at no cost from a yard sale when I mentioned I was a teacher and planned to use them in class.

Everyday items often found at yard sales can also be used in the science lab. If you have the need for a class set of mortars and pestles, but find them too expensive, search for metal spoons at yard sales. The back of a metal spoon works very well as a grinding tool (use a piece of wax paper to grind on). Hot plates are also often found at garage sales. I have six that I use and all were purchased at yard sales (none for more than a couple of dollars). Consider picking up plastic storage containers for storing your materials. I have found that square plastic food containers are great for

Figure 1

Science lesson plan websites

Discovery Education—
http://school.discovery.com/lessonplans/physci.html

Life science and biology lesson plans—
www.accessexcellence.org/AE/AEC/AEF

Educator's Reference Desk—
www.eduref.org/cgi-bin/lessons.cgi/Science

GoENC—
www.goenc.com

Academy Curricular Exchange—
http://ofcn.org/cyber.serv/academy/ace/sci/high.html

Science Lesson Plans and Resources—
www.cloudnet.com/~edrbsass/edsci.htm

Teachers First Lesson Plans—
www.teachersfirst.com/matrix.htm

LessonPlansPage—
www.lessonplanspage.com/ScienceJH.htm

holding small items such as marbles, balloons, and toy cars. Many larger storage bins with lids and handles can be found at these sales at bargain prices. Yard sales are unpredictable; you never know what you might discover. My prize find was a digital scale in great working condition that cost me only a dollar.

Other resources for science materials include local businesses such as hardware stores, film developing centers, newspaper offices, large corporations, medical clinics, doctors' offices, and bait shops. It is always a good idea to bring your school identification card or a letter from your school principal on school letterhead. My local hardware store offers discounts to teachers and free services. For example, when I needed a set of ramps, I asked my local hardware store to cut a set of eight 3 ft. long sections of 1 in. × 6 in. pine. They also routed out a groove in the center of each plank (for a marble to roll along) at no charge. We only had to pay the cost of the wood itself and that was at a discount. The entire set of eight ramps ended up costing less than $25. That is quite a savings over the $95 per ramp listed in my science supply catalog. In such cases, a thank-you card is also suggested.

There are numerous uses for film canisters in the science classroom. These can be obtained free of charge from most film development centers. One of these centers was located in a local grocery store. I asked them to start saving the canisters for me and they were more than happy to do so. Once a month, I stop by and pick up a large box filled with canisters that I share with the entire science department. A local newspaper office donates the blank end-rolls of newsprint to area teachers. These end-rolls are available in various widths and are great to use for many types of activities.

Many physicians are willing to donate anatomical and physiological posters and models and old medical journals to science teachers. Medical clinics are another resource for science materials. I obtained a set of old x-rays (with patients' identifying information removed) from one clinic that I used when I taught the human skeletal system. My classroom windows

Figure 2

Free and inexpensive science materials list

- Small cars (velocity)
- Marbles (velocity)
- Balloons (to show air has mass or capture yeast gas production)
- Metal spoons (mortars)
- Plastic containers (storage)
- Hot plates (heat)
- Pots (heat water)
- Aluminum pie pans (various uses)
- Magnifying glasses
- Balls (gravity and energy transfer)
- Heavy items such as fishing weights, large bolts, etc. (density)
- Tubs to hold water (density)
- Aquarium supplies
- Dropper bottles (pH experiments and biological stains)
- Craft supplies (for creative projects)
- Colored pencils, crayons, and markers (drawing activities)
- Rulers, compasses (rulers can be used as marble ramps)
- Magnets
- Simple machines (screwdrivers and tweezers)
- Flashlights (light and color)
- Baby-food jars and other small containers (volume)
- Science books
- Clay (to hold burning candles to observe chemical change)
- Candles (birthday and larger)
- Measuring cups and spoons
- Disposable cups and plates
- String
- Playing cards (probability lessons)
- Pot holders

worked as great light boxes when I taped the x-rays directly to them.

I found another source of materials in a local pulpwood mill that housed a chemical laboratory. They periodically donate a large supply of chemical laboratory glassware such as distillation tubes, specimen bottles, graduated cylinders, test tubes, and beakers. Of course, donated glassware must be thoroughly cleaned, sterilized, and inspected for cracks and chips before use by students.

I also frequently drop by the local bait shop to pick up night crawlers, crickets, and minnows to use in my biology classes. The local butcher is a great source for free animal organs, such as pig hearts, lungs, kidneys, and eyes, for in-class dissections.

One of the laboratory investigations that I use requires friction blocks, rectangular sections of wood with sandpaper attached to one side and an eyehook at one end. They cost $5–10 each when purchased from a science education supply catalog. Instead of buying friction blocks, I made my own set. I noticed that workers at a local construction site had amassed a large pile of wood blocks of various sizes and were preparing to burn them when I stopped by and asked if I could have some. I sanded down the rough edges,

purchased a set of eyehooks, and screwed one into an end of each block. Instead of gluing sandpaper to the blocks, I have students drag the blocks across sheets of sandpaper set on top of the lab table (one student will have to hold the sandpaper in place or it can be secured with masking tape).

Online auction houses such as eBay are another source of potentially cheap science materials. I recently purchased a dozen brand-new spring scales for $5. These scales sell for more than $6 each in science supply catalogs. I have also purchased microscope slides and coverslips, a model of a frog, and other items at steeply discounted prices from these online auction sites. At my school, we found it much cheaper to buy common household items used in laboratories (such as sodium chloride, ammonia, baking soda, cornstarch, and vinegar) in bulk from local discount warehouses rather than pay the prices charged by commercial supply houses.

Teaching science can be costly but it doesn't always have to be. Start your collection adventure now and you could soon be running out of storage space.

This article first appeared in the February 2007 issue of Science Scope.

Chapter 33

Got Stuff?

by Antonio M. Niro

When I began teaching many years ago, I was given a textbook that I estimate went into print sometime shortly after Gutenberg put monks out of business and just before Sputnik flew. Recently, a newly completed high school created the need for a new middle school, which in turn forced a new crisis—lots of students and no money for science supplies.

I learned quickly that science teachers have to be creative with limited funds to generate the equipment they need to accomplish hands-on science lab activities. Essentially, I had to "get stuff." Funds were lean during those times, but I needed ways and materials to teach heat, light, sound, magnetism, electricity, work, energy, and simple machines. I found great ideas for experiments in science books, sports, hobby magazines, journals, colleagues, conferences, the internet, and educational TV programs; however, I still needed the equipment to conduct hands-on activities.

So how do I get the stuff? To start, I asked students to bring in items; I went to local businesses; I pressed family and friends, but I did not generate enough, or the right kind of, materials. After brainstorming and rethinking my plan of attack, I decided to try a scavenger hunt. Students were given a list of materials to gather. I had created the list first from a textbook equipment list and then added to it based on a plan to conduct specific activities and experiments. Other items were added to the list while planning the materials that would be needed for future student demonstrations.

Because we were starting from scratch and generating materials for six teachers, my list was long—nearly 200 items. For your needs, consider including items that are consumables so budgeted school funds can be used to purchase permanent or expensive equipment that serve the whole school. Your list will vary based on the needs of your staff. In addition to the list, a letter was distributed for each student to take home a week before the deadline (see Figure 1, p. 158).

At a meeting of the parent-teacher organization, I explained what I had in mind and asked for their help, particularly with publicity and logistics. They got the word out through school notices, as well as the local newspaper and radio. In addition, they pledged a small financial commitment of about $50 to fund a pizza party for the winners. The approval of the administration was secured once the plan was finalized.

The rules of the hunt were simple: Students were given about one week to gather items; any amount of time longer than that fosters the "I've got plenty of time" attitude and may result in diminished returns. Point values were given to items depending on their rarity, availability, or desirability. Disposable items such as aluminum foil or string were assigned five

Figure 1

Scavenger hunt request letter

Dear Parents,

To assist the science teachers, your children have been asked to scour your basements, closets, garages, and attics in an effort to locate articles that will help us

- demonstrate concepts to students,
- provide additional materials used in student lab activities, and
- supplement classroom supplies or equipment.

Attached, you will find a list of items the science teachers are looking for, along with a point value. Each item or group of items is worth 5 points. Items with an asterisk (*) are worth 10 points.

Students will be asked to turn in materials on _____ to the organizers at school. Each student will be asked which homeroom they represent to keep track of the items returned. The homeroom that accumulates the greatest point total will be given a pizza party by the parent-teacher organization.

This is a great way to help out the science department, so pitch in if you can!

points. More desirable items such as tools or games were valued at 10 points. Permanent things such as an oscilloscope or a record player earned the greatest points. As items were collected, the students reported their grade and homeroom number so a tally of points could be maintained. With 24 homerooms participating, the homeroom with the greatest accumulated points won a pizza party. The party, which was held during a regular school lunch period, helped recognize the contribution of the parent-teacher organization and the efforts of the participants. The school's public address system may be incorporated to help build support during the collection week and at the conclusion of the hunt. The local newspaper, cable company, and school newsletter can also become important partners for this endeavor.

On a designated day, all of the students eagerly brought in everything they had collected. The items were counted, cataloged, and crated according to homeroom by parent volunteers who contributed their time to pick up and deliver. Without the help of parents, the collection of items could take place during science class or homeroom. Students are also quite able

to organize an effective force to help with collection and point tallying. Consider enlisting the support of staff to help gather items that are difficult for students to transport.

I got tons of stuff; in fact, the materials I received were only limited by the list that I had assigned. How surprised I was to learn of the community resources that I had never imagined! Items that were not even on the list came in as word got around that our school was finding ways to recycle all types of things. A really big surprise was some of the unexpected goodies that the community contributed, including

- a cutaway automobile battery (salesman's demo)
- several spools of wire (used for electrical experiments and electromagnets)
- oscilloscopes from a commercial lab that had upgraded (used to show characteristics of waves)
- expensive microscopes, boxes of glassware, and lenses, all from businesses that had also upgraded their equipment
- many examples of hand tools (used to demonstrate simple machines)
- a bicycle (used to illustrate gears and other machines)
- fish tubs (used as portable sinks for student experiments)
- small appliances (dissected to show electrical circuits and controls)
- old 45s (to make tops and gyroscopes)
- a tape recorder (used to capture sounds and echoes)

One mom showed up with a cardboard tube much like the ones found in a roll of paper towels. She asked me if I had a use for something like that and I said yes, I could find several uses for the tube—a pinhole camera, a telescope, a roller, or an arm for a mobile. I told her I could come up with a thousand ways to use them. "Good," she said, "that's just about how many I'll bring you." She came the following day with her minivan full of cardboard tubes. I became known as the "scavenger hunt guy" and was an ongoing recipient of equipment from local tech companies that were downsizing or upgrading equipment.

I received electronic equipment, including audio signal generators, incubators, and all types of animal cages and aquaria. What I could not use, I offered to others in our school district, and if no one could use it, I went to colleagues beyond our district. Materials were made available to science teachers first and then to others as well. Most materials were able to be stored in teacher classrooms, while larger items were held in the science department storage area or the custodian closet.

The scavenger hunt is a win-win-win situation. My students get access to materials that boost lab investigations. Parents have a chance to contribute and participate, as well as get rid of stuff. I get to spend more of my budget funds on permanent science equipment and not on consumables. A side benefit of a scavenger hunt is the eagerness of the students to see the items they contribute being used in science activities. Many times students have even suggested how materials might be used and brought them in to school. Often they showed more interest in the experiments that incorporated their stuff and contributed more because they developed "bragging rights." Try it. It works.

This article first appeared in the October 2003 issue of Science Scope.

Index

Page numbers in **boldface** type refer to tables or figures.